Root Cause Analysis

A Leader's Guide

Al Grainger

Copyright

For permission requests, please contact:
Al Grainger
grainger1@telus.net

Disclaimer:

This book is intended for informational purposes only. The author and publisher make no representations or warranties regarding the completeness, accuracy, or applicability of the content. The strategies and techniques presented are based on the author's experience and should be adapted to individual circumstances. The author and publisher disclaim any liability for the use or misuse of this information.

Second Edition Printed in Canada

Disclaimer

The examples and case studies presented in this book are drawn from real experiences and personal insights. To ensure privacy and confidentiality, all identifying details related to people, organizations, and specific situations have been altered or removed. Any resemblance to actual persons, organizations, or events is purely coincidental. The purpose of these examples is to illustrate key concepts and provide practical insights based on real-world applications. Readers should use their own judgment and consider their unique contexts when applying the principles and practices discussed in this book.

Dedication

Nothing in my life would be possible without the support from the love of my life and my best friend, Shani, and our children, Ben and Matt.

Foreword

Root Cause Analysis (RCA) is more than just a methodology; it is a philosophy of problem-solving that has the power to transform how organizations operate. In this book, Al Grainger has distilled decades of expertise and experience into an accessible, practical guide that transcends industries and business sizes. Whether you are leading a Fortune 500 company or a small, growing enterprise, the principles and tools outlined in this book offer a roadmap for sustainable improvement and innovation.

Al's ability to demystify complex processes and present them in a clear, actionable way is evident throughout these pages. He doesn't just explain the "what" of RCA but dives deep into the "how" and "why," empowering readers to go beyond treating symptoms and instead tackle the root causes of challenges they face. The emphasis on collaboration, structured problem-solving, and fostering a culture of accountability positions this book as an invaluable resource for leaders and teams alike.

As you explore the case studies, tools, and strategies presented here, you'll find a wealth of actionable insights that will enhance your problem-solving capabilities. Al's pragmatic, down-to-earth approach makes this book not just a learning resource but a practical companion to drive meaningful change in your organization.

I encourage you to immerse yourself in this work, apply its lessons, and embrace the mindset it fosters. The rewards—operational excellence, engaged teams, and a resilient organization—are well worth the effort.

Joseph R. LaCount
Global Quality and Continuous Improvement Leader, Lean Six Sigma Master Black Belt.

Preface

My original goal was to create a practical guide to Root Cause Analysis (RCA), helping leaders solve problems at their source instead of wasting time on temporary fixes. But as I wrote, I kept diving deeper into Lean principles, realizing that many organizations struggle not just with identifying root causes, but with the fundamental ways they approach improvement.

At some point, it became clear: I wasn't writing one book—I was writing two. The Lean concepts, while essential, needed their own space. So, I split the manuscript, refining my focus. That decision led to ***Lean Thinking for Leaders: A Beginner's Blueprint***, a practical introduction to Lean principles for business leaders who want to drive efficiency and improvement without relying on expensive consultants.

This book, ***Root Cause Analysis: A Leader's Guide,*** is the book I originally set out to write. It tackles one of the biggest challenges I've seen across industries: organizations repeatedly solving the wrong problems. Without understanding the real root cause, teams end up treating symptoms rather than fixing the underlying issue, wasting time, money, and effort. My goal is to give leaders the tools to break that cycle, create lasting solutions, and build a culture of continuous improvement.

I hope you find this book practical, actionable, and most importantly, helpful in tackling the challenges that stand in the way of real progress.

Al Grainger

Table Of Contents

Introduction: The Importance of Root Cause Analysis

Root cause analysis (RCA) is an essential methodology for effective problem-solving and operational excellence across all sectors. By uncovering and addressing the fundamental causes of issues, rather than just treating symptoms, organizations can achieve sustainable improvements and prevent recurring problems. RCA provides a structured approach to identifying underlying issues and implementing solutions that address the root causes, leading to more lasting and effective results.

In today's competitive business environment, reactive problem-solving is no longer sufficient. Organizations that rely on quick fixes to meet short-term goals often find themselves facing the same issues repeatedly, resulting in wasted resources, inefficiencies, and diminished employee morale. For business leaders, adopting RCA is not only about resolving immediate problems but also about fostering a culture of continuous improvement - a critical aspect of maintaining competitive advantage.

RCA is not just for manufacturing. This methodology is applicable to any business or industry where processes fail to meet objectives, from safety incidents to product failures, to customer service issues. Whether you are in healthcare, finance, or the public sector, RCA can help identify the underlying causes of problems and ensure that they do not recur. Many organizations face similar challenges - whether it is a production delay in a factory or a compliance failure in a hospital - the principles of RCA remain relevant.

A Pragmatic, Down-to-Earth Approach

This book is designed as a pragmatic, down-to-earth guide to help business leaders apply RCA effectively. You do not need to be an expert in complex methodologies or technical jargon to understand and implement RCA. This is about using straightforward, practical tools to get to the heart of the problem and develop solutions that last.

Throughout the chapters, we will provide real-world examples, easy-to-follow steps, and hands-on tools to guide you through the RCA process, so you can see results in your own organization. This guide is as useful for those leading large teams with major production issues as it is for small businesses or departments managing simple, time-consuming concerns. RCA is not about theory; it is about solving real problems that impact your bottom line and your people.

RCA's Strategic Role in Organizations

Beyond problem-solving, RCA plays a strategic role in shaping an organization's future. Business leaders who implement RCA effectively can identify areas for innovation, optimize operations, and reduce waste, contributing to broader organizational goals. RCA aligns with methodologies like Lean Manufacturing and Six Sigma, providing a foundation for continuous improvement that resonates across departments, from operations and production to human resources and customer service.

For example, RCA has helped companies like Toyota and GE maintain their positions as industry leaders by embedding root cause analysis into their daily operations. By consistently applying RCA principles, these companies avoid the costly impact of recurring issues and improve overall productivity.

Employees are a crucial component of any organization and often overlooked when issues arise. When problems persist without effective resolution, it can lead to decreased morale and engagement. Employees can feel like they are being blamed for issues that are often process-related. Studies show that employee-related causes for production issues are significantly less common than process-related or reliability issues. Involving employees in the RCA process helps address the root causes while simultaneously boosting engagement and morale. Employees who are included in problem-solving feel valued and are more likely to be motivated, leading to improved job satisfaction and performance.

What You Will Learn in This Book

This book offers a comprehensive guide to RCA, providing a step-by-step approach that is practical for business leaders and operational managers. While the application of the learnings may seem focused on manufacturing, these techniques apply across industries, from healthcare to finance, public sector, and beyond. Each chapter will break down the essential components of RCA, offering tools and techniques that leaders can use to implement RCA in their own organizations.

Readers will learn:

- How to identify the root causes of problems and avoid superficial fixes.
- The importance of documenting RCA outcomes for long-term reference and accountability.

- How to use the A3 problem-solving template and other RCA tools like 5 Whys and Fishbone Diagrams.
- Best practices for running RCA sessions and encouraging cross-functional collaboration.
- How to quantify potential solutions, shortlist the most effective ones, and create an action plan for implementation.
- Strategies for ensuring the long-term success of RCA solutions by tracking progress and using methods like Plan-Do-Check-Act (PDCA) to sustain improvements.

The Value of RCA for Business Leaders

RCA is more than just a problem-solving tool—it's a critical driver of operational excellence and strategic decision-making. Business leaders who master RCA can prevent costly disruptions, improve employee engagement, and drive continuous improvement throughout their organizations. Whether readers are new to RCA or looking to refine their existing practices, this book is designed to equip them with the knowledge and tools they need to drive meaningful, lasting improvements in their organizations.

Through Al Grainger's experience and expertise, readers will gain a deeper understanding of why pinpointing the root cause of problems is crucial for achieving success and sustaining excellence in any industry. The following chapters will explore RCA's applications in real-world scenarios, illustrating how identifying the root cause can be the difference between temporary solutions and long-term success.

Chapter 1: The Foundation of Root Cause Analysis

Introduction to Root Cause Analysis (RCA)

Root Cause Analysis (RCA) is a systematic, evidence-based approach to identifying and addressing the underlying causes of problems, rather than simply treating symptoms. By focusing on the fundamental causes, organizations can develop long-lasting solutions, leading to operational excellence, cost savings, and improved customer satisfaction. RCA is applicable across all sectors—whether in manufacturing, healthcare, IT, finance, or public administration—making it an essential tool for leaders looking to improve their processes, products, and services.

Unlike quick fixes or band-aid solutions, RCA takes a holistic approach to problem-solving. This process often begins with a thorough investigation into contributing factors, followed by analyzing how and why an issue occurred, and finally, creating actionable steps to resolve it at its core. As a result, organizations not only solve the immediate problem but also prevent recurrence, improve efficiency, and foster a culture of continuous improvement.

Why RCA Matters

Root Cause Analysis is more than just a problem-solving tool; it is a strategic approach that fosters long-term success. For organizations, RCA offers a wide range of benefits:

1. **Preventing Recurrence**:
   RCA addresses the root causes of issues, ensuring that problems do not resurface. Without RCA, organizations often experience recurring issues, leading to wasted time and resources. For example, Toyota's early adoption of RCA—often through its "5 Whys" technique—helped reduce assembly line issues that would have caused repeated production delays.

2. **Increasing Efficiency**:
   By identifying the underlying causes of inefficiencies, RCA helps organizations streamline workflows and remove obstacles that slow down operations. This process leads to improved productivity, lower operational costs, and reduced downtime. A study conducted by Deloitte found that RCA, when applied to manufacturing and production, led to a 15% increase in efficiency by addressing bottlenecks at their root.

3. **Improving Quality**:
   RCA plays a critical role in quality control by uncovering the real sources of defects or process failures. By applying RCA techniques, companies can enhance the consistency of their products or services. In the healthcare sector, for example, root cause analysis is widely used to identify the root causes of medical errors, leading to better patient outcomes and reduced liability.

4. **Reducing Costs**:
   Effective RCA can significantly reduce costs associated with recurring problems, inferior quality, or unplanned downtime. Costs linked to misdiagnosing issues, fixing the wrong problems, or addressing symptoms instead of the root causes are often substantial. By adopting RCA, organizations can cut down on unnecessary repairs, lost revenue, and operational inefficiencies.

5. **Customer Satisfaction**:
   Improve customer satisfaction by addressing root causes of issues that impact product or service quality, thereby avoiding dissatisfaction and enhancing customer loyalty.

For instance, in manufacturing, an unresolved quality issue might lead to repeated defects and rework, resulting in increased costs, customer dissatisfaction, and reduced employee morale. RCA provides a framework to understand and resolve such issues comprehensively, leading to improved product quality, operational efficiency, and cost savings, while also enhancing employee morale by addressing the root causes of persistent issues.

Employee Satisfaction and Engagement

Root cause analysis is not just about process improvement; it is about engaging employees in meaningful problem-solving efforts. Employees, when included in the RCA process, feel empowered to contribute their insights and expertise, which can lead to better outcomes and higher job satisfaction. Research suggests that employee engagement rises significantly when workers are part of cross-functional RCA teams, as it gives them ownership over solving persistent issues.

Involving employees helps improve morale, as they feel valued when they see their input led to real changes in the organization. Additionally, addressing root causes can alleviate common frustrations that arise from dealing with repetitive or unresolved problems, leading to higher retention rates and improved performance across the board.

Common Pitfalls in Problem-Solving

Organizations often fall into the trap of addressing surface-level symptoms rather than tackling the real problems beneath them. These include:

1. **Quick Fixes**:
   Companies often implement temporary solutions that only address the symptoms of a problem rather than its root causes. While quick fixes may offer short-term relief, the underlying issue often resurfaces, leading to a cycle of recurring problems. A 2017 study found that organizations relying on "band-aid" fixes experienced 30% higher operational costs due to the ongoing nature of unresolved issues.

2. **Jumping to Conclusions**:
   Many organizations are eager to fix problems without conducting a thorough investigation. Without fully understanding the root cause, these efforts often lead to the wrong solutions being implemented. This approach wastes resources and may even exacerbate the original issue.

3. **Lack of Comprehensive Investigation**:
   Failing to investigate thoroughly can prevent organizations from identifying the root cause. A proper RCA requires gathering data, analyzing multiple contributing factors, and examining all potential causes before determining the best course of action.

Addressing these pitfalls requires adopting a structured RCA process that emphasizes patient, data-driven analysis over hasty decision-making.

The RCA Process Overview

A successful root cause analysis involves a structured process designed to ensure a thorough investigation and lasting resolution. These steps are:

1. **Identify the Problem**:
   Define the problem clearly and collect relevant data to understand its scope. Use metrics to quantify the issue, such as production downtime or customer complaints.

2. **Analyze the Problem**:
   Apply tools like the 5 Whys, Fishbone Diagrams, or Pareto Analysis to identify the root causes. These methods help uncover systemic issues that might not be immediately apparent.

3. **Develop Solutions**:
   Create practical, sustainable solutions that address the root cause of the problem. Ensure that proposed solutions are realistic and feasible.

4. **Implement Solutions**:
   Implement the chosen solution in a structured way, ensuring all team members understand their roles and responsibilities. Use change management strategies to support this step.

5. **Evaluate Results**:
   Monitor the effectiveness of the solutions and assess whether the problem has been resolved. Gather feedback and make any necessary adjustments. Document lessons learned to improve future RCA efforts.

Case Study: Addressing Recurring Downtime in Manufacturing

In a manufacturing company, recurring downtime led to significant production delays, costing the company hundreds of thousands in lost revenue each year. Despite numerous efforts to reduce downtime, the problem persisted.

RCA Process:

- **Identifying the Problem**: The team identified that unplanned machinery downtime was the primary cause of production delays.
- **Analyzing the Problem**: Using techniques such as the 5 Whys, the team discovered that maintenance was reactive rather than preventive, leading to frequent equipment breakdowns.
- **Developing Solutions**: The solution involved implementing a preventive maintenance schedule and investing in better training for machine operators.
- **Implementing the Solution**: Over a period of six months, the company rolled out preventive maintenance across all departments.
- **Evaluating Results**: Within the first quarter, machinery downtime reduced by 40%, leading to an estimated $500,000 in savings.

This case demonstrates how RCA can identify deep-rooted issues that are not immediately obvious and provide long-lasting solutions.

Summary and Key Takeaways

This chapter has introduced the concept of Root Cause Analysis and highlighted its importance in achieving operational excellence. Key points discussed include:

- The need to address root causes rather than symptoms.
- The benefits of RCA, including preventing recurrence, increasing efficiency, improving quality, and reducing costs.
- Common pitfalls in problem-solving and the importance of thorough investigation.
- An overview of the RCA process and its steps.

Understanding and implementing RCA is crucial for professionals in any industry who seek to drive meaningful improvements and achieve sustainable success.

Chapter 2: How to Document a Root Cause Analysis

Root Cause Analysis (RCA) is a structured process that helps organizations identify and address the underlying causes of problems, ensuring that they achieve long-term solutions rather than temporary fixes. However, conducting an RCA is only part of the process. Equally important is the task of documenting the findings and outcomes. Proper documentation plays a vital role in ensuring that RCA results are more than just fleeting insights—they become lasting contributions to the organization's knowledge base.

Documentation creates a reliable reference point for future problem-solving and helps organizations build on past lessons to avoid repeating mistakes. It also supports continuous improvement, enabling businesses to leverage these insights when solutions do not meet expectations. This chapter delves into why documenting RCA outcomes is essential, how it supports organizational learning, and how businesses can effectively utilize problem-solving templates to capture RCA results.

Enhancing Organizational Learning

Documenting RCA outcomes is not merely about creating a record of solved problems—it's about building a living knowledge repository that continues to benefit the organization over time. This knowledge base becomes a cornerstone for employee development, cross-departmental learning, and long-term improvement.

Training and Development

New employees often face a steep learning curve when joining an organization. RCA documentation helps bridge this gap by providing them with insights into the company's problem-solving methods, without relying solely on direct experience. These documented RCA cases offer practical, real-world examples of how the organization approaches complex issues, offering new hires a wealth of knowledge to draw from as they grow into their roles.

For instance, if recurring equipment failures were addressed using an RCA, the documented solution can serve as a guide for future employees. The documentation not only covers the technical fix but also explains the rationale behind the actions taken. Smith (2018) highlights that such documentation allows new hires to understand the decision-making processes that underpin the company's operations. This speeds up learning, reduces mistakes, and fosters a self-sufficient workforce capable of tackling similar challenges with confidence.

Continuous Improvement

Another significant benefit of RCA documentation is its contribution to continuous improvement. Organizations that maintain thorough documentation of past RCAs can review these cases to identify patterns or recurring issues. This allows them to not only address isolated incidents but also uncover systemic problems that may require more comprehensive solutions.

By regularly reviewing documented RCAs, leadership teams can spot trends in operational inefficiencies, safety hazards, or quality control issues. Over time, this practice helps foster a culture of continuous improvement, where lessons learned from one department can be shared and applied across the organization. Shared Root Cause Analysis documentation becomes a tool for innovation, driving initiative-taking rather than reactive organizational development.

Supporting Accountability and Transparency

In any organization, accountability and transparency are key elements of success. RCA documentation plays a pivotal role in ensuring that all parties involved are clear on the findings, assigned actions, and steps required to resolve issues.

Assigning Responsibility

One of the main functions of RCA documentation is to clearly assign ownership for corrective actions. Without proper documentation, it's easy for tasks to fall through the cracks, leading to delays or incomplete solutions. By documenting who is responsible for each action, organizations create accountability and ensure that all necessary steps are followed through. This is especially crucial when multiple departments or teams participate in the RCA process.

For instance, the production team may be responsible for modifying a specific process, while the quality control team monitors the results. By documenting these responsibilities, RCA ensures that there is no confusion about who oversees each task. Brown (2019) notes that when RCA documentation explicitly outlines responsibilities, it reduces the risk of miscommunication and increases the likelihood of timely follow-through.

Internal Transparency

Transparency is not just about assigning roles—it's about making sure that the entire organization is on the same page. This is particularly important when dealing with safety issues or any situation where a failure could have dire consequences. RCA documentation provides transparency for internal stakeholders by clearly communicating the risks, the root causes identified, and the corrective actions being implemented.

When safety concerns arise, being transparent about the RCA process builds trust and ensures that employees feel confident that the organization is addressing the problem in a thorough, responsible manner. Experience shows that transparency around RCA outcomes promotes a culture of openness, where employees understand the steps being taken to protect their well-being and ensure compliance with safety standards.

Reference Point for Solution Failures

Even with the most well-executed RCA and corrective action plans, there are times when solutions do not deliver the expected results. RCA documentation serves as a critical reference point for reassessing and adjusting strategies when things do not go as planned.

Learning from Failure

Failure is an inevitable part of innovation and growth. When a solution falls short, RCA documentation provides a comprehensive review of the original problem, the proposed solutions, and the reasoning behind them. This allows teams to revisit the issue with fresh perspectives and avoid retracing steps that didn't work.

In many cases, RCA sessions identify multiple potential solutions, but not all can be implemented immediately due to resource constraints or other limitations. When organizations document these unimplemented solutions, they create a reference point for future problem-solving. If the original solution fails, the organization can turn to the documented alternatives, saving time and effort in developing new strategies.

For instance, if a new maintenance schedule was implemented following an RCA but equipment breakdowns persist, the original documentation can be reviewed, and other potential solutions can be implemented. Failing to retain record from your RCA sessions means you must redo the process. Further, it devalues the efforts from the original session and reduces employee engagement.

Facilitating Communication and Collaboration

RCA documentation is a powerful tool for improving communication across departments and fostering collaboration. When different departments contribute to solving a complex problem, proper documentation ensures that everyone has a clear understanding of the findings, decisions, and follow-up actions.

Cross-Functional Collaboration

In many industries, particularly in manufacturing or healthcare, solving a complex problem requires input from multiple departments. For instance, a production issue might involve input from the production team, maintenance team, and quality control. When RCA documentation is shared across these departments, it ensures that everyone is working with the same information, and each department's expertise will be applied to develop a comprehensive solution.

Let's say a production bottleneck is causing delays. By involving multiple departments in the RCA process and sharing the documentation, the organization can identify potential solutions from different perspectives. The bottleneck could be related to equipment maintenance, inventory management, or workflow design. By sharing insights through RCA documentation, each department can contribute to a more effective, holistic solution.

Engaging External Stakeholders

RCA documentation is not just for internal teams—it can also be shared with external stakeholders like suppliers, contractors, or even customers. When external parties are part of the problem or solution, sharing RCA documentation can improve alignment and foster stronger collaboration. For instance, if a supplier consistently delivers substandard materials, involving them in the RCA process and sharing the outcomes can help them understand the impact of their materials on your organization's operations. This can lead to better quality control and stronger relationships moving forward (Roberts & Wilson, 2018).

Ensuring Regulatory Compliance

Many industries are governed by strict regulations that require thorough documentation of processes, including RCA outcomes. Whether for legal compliance or regulatory audits, RCA documentation plays a critical role in demonstrating that an organization has taken appropriate corrective actions.

Legal Compliance and Audits

For industries such as healthcare, finance, or manufacturing, documenting RCA outcomes is not just a best practice—it is often a legal requirement. Regulatory bodies may require detailed records of how issues were identified, the root causes, and the steps taken to resolve them. For example, in the healthcare sector, hospitals must document RCA processes following adverse patient events to demonstrate that the organization took the necessary steps to prevent future occurrences.

This documentation is also invaluable during audits. Whether it is a routine quality audit or a regulatory inspection, having detailed records of the RCA process makes the audit process smoother and ensures that the organization can provide evidence of its problem-solving efforts. RCA documentation not only ensures compliance but also serves as a key defense in legal disputes, demonstrating that the organization acted responsibly to mitigate risks.

Chapter 3: Problem Solving Tools

Using structured problem-solving templates during RCA sessions provides clarity, consistency, and focus. Templates ensure that the RCA process is well-documented and that all relevant aspects of the problem and its solution are captured. Here are three commonly used templates:

5 Why

The 5 Whys is a simple but powerful technique used to explore the root cause of a problem by repeatedly asking the question "Why?" The method is part of the lean manufacturing philosophy and aims to identify the underlying causes of issues to prevent recurrence. Here's a detailed guide on how to use the 5 Whys technique:

Step-by-Step Guide to Using the 5 Whys

1. Define the Problem

- Identify the Problem: Clearly state the problem you want to investigate. It should be a specific and observable issue.
- Document the Problem Statement: Write down the problem statement to ensure everyone involved has a clear understanding of what is being analyzed.

2. Form a Team

Assemble a Cross-Functional Team: Include individuals who are familiar with the process or system in question. Diverse perspectives help uncover various aspects of the problem.

3. Ask the First "Why?"

- Identify the Immediate Cause: Ask the team why the problem is occurring. The answer should address the immediate or most obvious cause.
- Document the Answer: Write down the first answer. This forms the basis for the next question.

4. Ask "Why?" Again

- Drill Down Further: Take the answer to the first question and ask why that cause is happening. This helps dig deeper into the process.
- Document the Answer: Write down the second answer. This process continues until you have asked "Why?" five times.

5. Continue Asking "Why?"

- Repeat the Process: Continue asking "Why?" for each subsequent answer. The goal is to reach the root cause, which may take fewer or more than five iterations.
- Validate Each Step: Ensure that each answer is logical and based on factual data. Avoid jumping to conclusions without evidence.

6. Identify the Root Cause

- Reach a Consensus: Once the team agrees that they have identified the root cause, document it clearly.
- Verify the Root Cause: Ensure that the identified root cause is the true underlying reason for the problem. It should address the issue in a way that prevents recurrence.

Example of the 5 Whys in Action

Problem: The Manufacturing Line Stopped		
	Question	**Answer**
Why # 1	Why did the manufacturing line stop?	The conveyor belt stopped moving.
Why # 2	Why did the conveyor belt stop moving?	The drive motor stopped running.
Why # 3	Why did the drive motor stop running?	The motor overheated.
Why # 4	Why did the motor overheat?	There was insufficient lubrication in the motor bearings.
Why # 5	Why was there insufficient lubrication in the motor bearings?	The lubrication schedule was not followed.
Root Cause: The lubrication schedule was not followed.		

Solution: Implement a stricter maintenance schedule and ensure adherence through regular audits and training for the maintenance staff.

Tips for Effective Use

- **Stay Focused:** Ensure each "Why?" is related to the previous answer. Avoid straying into unrelated issues.
- **Be Specific:** Provide clear and concise answers to each "Why?" question. Vague responses can obscure the root cause.
- **Involve the Right People:** Include team members with direct knowledge of the process or system being analyzed.
- **Use Data:** Base each answer on factual data and observations rather than assumptions or guesses.
- **Iterate as Needed:** While the name suggests five iterations, you may need more or fewer to reach the true root cause.

The 5 Whys technique helps teams delve deep into problems to identify and address root causes, leading to more effective and lasting solutions.

Fishbone Diagram

A fishbone diagram, also known as an Ishikawa diagram or cause-and-effect diagram, is a tool used for identifying and analyzing the root causes of a problem. The Fishbone Diagram organizes potential causes of a problem into categories such as people, materials, environment, processes, equipment, and measurements. This method is especially useful for complex problems with multiple contributing factors. The diagram helps teams systematically explore different areas that may be causing the issue, visually highlighting potential impacts leading to the underlying issue.

Here is a detailed guide on how to use it:

Step-by-Step Guide to Using a Fishbone Diagram

1. Define the Problem Statement

- **Identify the Problem:** Clearly state the problem or effect you are trying to analyze. This should be a specific issue that you want to investigate.
- **Write the Problem Statement:** Write the problem statement at the head of the fishbone diagram, usually on the right-hand side.

2. Draw the Main Diagram

- **Draw the Spine:** Draw a horizontal line across the page. This represents the spine or backbone of the fish.
- **Add the Head:** At the right end of the spine, draw a box or circle to enclose the problem statement. This is the "head" of the fish.

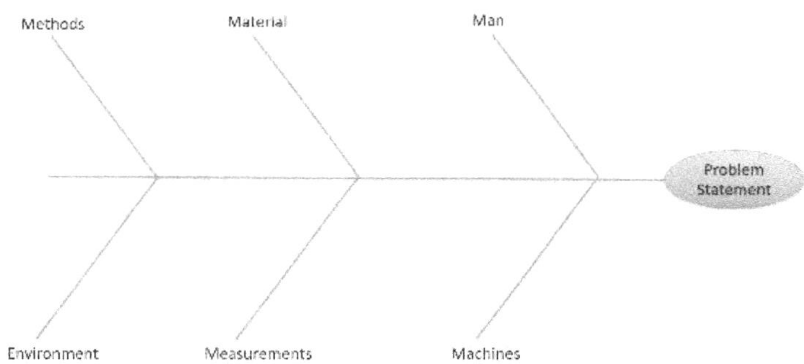

3. Identify Major Categories of Causes

- **Determine Categories:** Identify the major categories of potential causes. Common categories in manufacturing include:
    - Methods: Processes and procedures used.
    - Machines: Equipment and tools.
    - Materials: Raw materials and consumables.
    - Workforce: People involved in the process.
    - Measurements: Data and metrics used to measure performance.
    - Environment: Physical conditions and external factors.
- **Draw the Categories:** Draw branches off the spine for each category. These branches should angle out from the spine like the bones of a fish.

4. Brainstorm Possible Causes

- **Sub-Causes:** For each major category, brainstorm potential causes that could contribute to the problem. These are sub-causes.

- **Add Sub-Causes:** Draw smaller branches off each major category branch to represent these sub-causes. Keep asking "why" to dig deeper and identify root causes. This process is sometimes called the 5 Whys, where you ask "why" up to five times to drill down to the underlying cause.

5. Analyze and Discuss

- **Examine the Diagram:** Review the fishbone diagram to identify patterns or common themes among the causes.
- **Prioritize Causes:** Determine which causes are most likely to be the root causes of the problem. This might involve further investigation, data collection, or discussion with team members.
- **Validate Causes:** Ensure that the identified causes are valid and supported by evidence. This may involve testing hypotheses or collecting additional data.

6. Develop Action Plan

- **Create Solutions:** Based on the identified root causes, develop solutions or corrective actions to address the problem.
- **Implement Solutions:** Put the solutions into action. Assign responsibilities and timelines to ensure accountability.
- **Monitor Results:** Track the effectiveness of the solutions and adjust, as necessary.

Example of a Fishbone Diagram in Action

Imagine a manufacturing company experiencing a high rate of defective products. The team might create a fishbone diagram with the problem statement "High Rate of Defective Products" at the head. The major categories could be Methods, Machines, Materials, Manpower, Measurements, and Environment. For each category, the team would brainstorm potential causes and add them as sub-branches. For example, under Methods, they might list "incorrect assembly procedure" or "lack of standard operating procedures." Under Machines, they might list "machine calibration issues" or "equipment wear and tear."

Tips for Effective Use

- **Collaborative Effort:** Involve team members from different departments and levels of the organization to get diverse perspectives.
- **Visual Clarity:** Ensure the diagram is clear and easy to read. Use assorted colors or shapes to distinguish between categories and sub-causes.
- **Focus on Root Causes:** Avoid listing symptoms of the problem. Aim to identify underlying causes that, if addressed, would prevent the problem from recurring.
- **Iterative Process:** The fishbone diagram can be revisited and updated as more information becomes available or as the problem evolves.

Using a fishbone diagram helps teams systematically explore potential causes of a problem, leading to more effective solutions and improvements.

Pareto

A Pareto chart is a graphical tool used to prioritize problems or causes by highlighting the most significant factors in a dataset. It is based on the Pareto Principle, also known as the 80/20 rule, which states that 80% of the effects come from 20% of the causes. Here's how to effectively use a Pareto chart:

Step-by-Step Guide to Using a Pareto Chart

1. Collect and Organize Data

- **Identify the Problem Area:** Determine the specific problem or process you want to analyze.
- **Collect Data:** Gather data related to the problem. This could be defects, customer complaints, production delays, etc.
- **Categorize Data:** Group the data into categories or causes. For example, if analyzing defects, categories might include "scratches," "missing parts," "misalignments," etc.

2. Create a Frequency Table

- **List Categories:** List all the categories of causes in descending order of frequency (or cost, time, etc.).
- **Calculate Frequencies:** Count the number of occurrences for each category.
- **Calculate Cumulative Frequencies:** Calculate the cumulative frequency for each category, which shows the running total of occurrences.

3. Construct the Pareto Chart

- **Draw the Axes:** On a graph, draw two vertical axes. The left vertical axis represents the frequency of occurrences, and the right vertical axis represents the cumulative percentage. The horizontal axis represents the distinct categories of causes.

- **Create Bars:** Draw bars for each category. The height of each bar corresponds to its frequency. Arrange the bars in descending order from left to right.
- **Plot Cumulative Line:** Starting from the left, plot a line graph showing the cumulative percentage of occurrences for each category. This line starts at the top of the first bar and ends at 100% on the rightmost category.

4. Analyze the Chart

- **Identify Vital Few:** Look for the categories that make up approximately 80% of the cumulative frequency. These are typically the first few bars on the left.
- **Focus on Major Causes:** Concentrate your efforts on addressing these key categories, as they have the most significant impact on the problem.

5. Implement Improvements

- **Develop Action Plans:** Create action plans to address the major causes identified in the Pareto chart.
- **Monitor Changes:** Implement the improvements and monitor their effects on the problem.
- **Reevaluate Regularly:** Regularly update the Pareto chart with new data to ensure the improvements are effective and to identify any new significant causes.

Example of a Pareto Chart in Action

Problem: High number of customer complaints in a service company.

Data Collection:

Categories of complaints: "Long Wait Time," "Rude Staff," "Billing Errors," "Service Errors," "Poor Communication."

Frequency Table:

- Long Wait Time: 50 complaints
- Rude Staff: 30 complaints
- Billing Errors: 20 complaints
- Service Errors: 15 complaints
- Poor Communication: 10 complaints

Cumulative Frequency:

  - Long Wait Time: 50
  - Rude Staff: 80 (50+30)
  - Billing Errors: 100 (80+20
  - Service Errors: 115 (100+15)
  - Poor Communication: 125 (115+10)

Construct the Chart

- Draw bars for each category in descending order: Long Wait Time, Rude Staff, Billing Errors, Service Errors, Poor Communication.
- Plot the cumulative percentage line.

Analysis:

- The chart shows that "Long Wait Time" and "Rude Staff" together account for 80% of the complaints.
- Focus on reducing wait times and improving staff behavior.

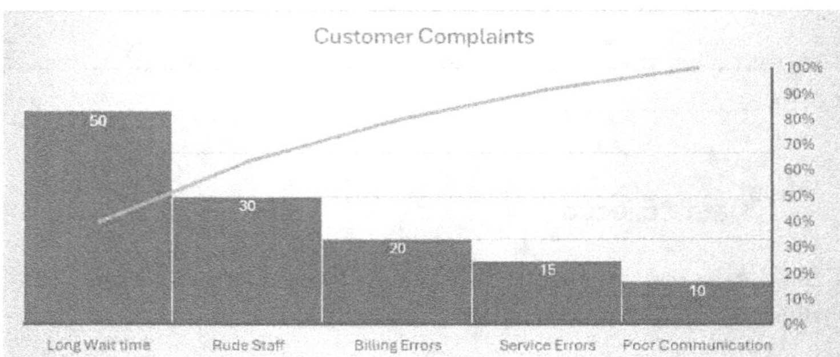

Tips for Effective Use

- **Accurate Data Collection:** Ensure the data collected is accurate and relevant to the problem.
- **Clear Categories:** Clearly define categories to avoid overlap and confusion.
- **Visual Clarity:** Make the chart easy to read by clearly labeling axes and categories.
- **Use in Combination:** Combine Pareto charts with other tools like the 5 Whys or Fishbone Diagrams for a deeper analysis.

Using a Pareto chart helps prioritize efforts on the most significant causes of a problem, leading to more effective and efficient solutions.

A3 Template

Originating from Lean manufacturing, the A3 template condenses the entire problem-solving process onto a single sheet of paper. This format encourages clear, concise documentation, ensuring that the problem, root causes, and solutions are communicated effectively. Davis (2021) notes that the A3 template is particularly valuable in organizations that prioritize clarity and efficiency in their RCA processes.

Chapter 4: The A3 Template

History and Origin of the A3 Template

The A3 problem-solving template was developed as part of the Lean manufacturing philosophy in Toyota Motor Corporation. The name "A3" refers to the paper size used to document the process. The A3 paper size is 11.69 by 16.54 inches or in metric 30 x 42 centimetres. The North American equivalent is the tabloid size, 11 x 17 inches. However, the significance of the A3 template transcends its physical dimensions. At Toyota, the A3 became a cornerstone of their continuous improvement culture, embodying the company's commitment to systematic problem-solving, collaboration, and clear communication.

Toyota designed the A3 template to address problems in a structured manner while fostering transparency and accountability. By distilling complex issues into a single-page document, this visual tool ensured that everyone, from assembly line workers to executives, could understand and contribute to the problem-solving process. The A3 was not merely an instrument, but a symbol of Toyota's dedication to operational excellence.

How it is Used:

The A3 template is designed to guide teams through a systematic approach to problem-solving. It acts as both a roadmap and a record of the process, ensuring that all aspects of a problem are considered and documented. The A3 process is widely used across various industries to address operational inefficiencies, quality issues, or other organizational challenges.

The main goal of the A3 template is to simplify complex problems by distilling them into a visual, easy-to-understand format. This structure is particularly useful in large organizations where cross-departmental communication is necessary. The single-page format forces teams to concise their analysis and focus only on the most relevant aspects of the problem and its solution, cutting through unnecessary detail and allowing for quick review and action by leadership and stakeholders.

To What End?

The A3 template is not just about solving immediate problems; it is used to cultivate a culture of learning and continuous improvement. The structured approach ensures that root causes are identified, the solutions are thoughtfully implemented, and long-term sustainability is considered. The A3 template prevents teams from rushing into "band-aid" fixes by requiring a thorough examination of the problem.

- **Decision-Making:** Once completed, the A3 report becomes a powerful tool for decision-makers. Its clear, organized structure provides all the necessary information at a glance, enabling managers and executives to make informed decisions without having to sift through lengthy reports.
- **Communication and Collaboration:** The A3 document is often shared across different departments to foster collaboration. For instance, a problem affecting production might also impact the supply chain, so involving both departments in the A3 process ensures that the solution addresses all related issues.
- **Accountability:** The A3 template serves as a record of accountability. Since each section assigns specific responsibilities, the A3 holds individuals and teams responsible for executing the countermeasures and ensuring that the results are achieved.

What Happens with It?

Once an A3 report has been completed, it does not just sit in a filing cabinet—it is a living document. The A3 report is often updated as the situation evolves, or as new data becomes available. Here is what typically happens next:

1. **Presentation and Review:**
   - Once completed, the A3 report is presented to senior management or other stakeholders for review. The team walks through the problem, the analysis, and the proposed solutions. This is an opportunity for feedback and for ensuring that the solution aligns with the organization's strategic goals.
   - Some organizations share the A3 publicly within the company to foster transparency and cross-functional learning. Teams can learn from each other's problem-solving processes by reviewing A3 reports, contributing to a broader **organizational memory.**
2. **Implementation and Monitoring:**
   - After approval, the solutions outlined in the A3 template are implemented. The document remains a point of reference throughout the implementation phase, ensuring that the team stays on track with deadlines and responsibilities.
   - Teams monitor the effectiveness of the implemented solutions, referring to the A3 to confirm whether the anticipated outcomes were achieved. If necessary, the A3 is updated to reflect any adjustments made during implementation.

3. **Continual Use for Continuous Improvement:**
    - Even after the problem has been resolved, the A3 report remains a valuable reference. It can be used for training new employees, as a case study in process improvement discussions, or as a basis for future problem-solving efforts. It may also be revisited during periodic reviews to ensure that the solution continues to hold up over time.
    - The A3 is often stored as part of the company's knowledge management system, where other teams can access it to prevent similar issues from occurring in the future.

The A3 template is much more than just a tool for solving problems in the moment. It plays a critical role in guiding decision-making, fostering cross-departmental collaboration, and ensuring accountability at all levels. Its legacy within an organization often extends beyond the immediate problem, as it serves as a record of lessons learned and a foundation for continuous improvement efforts.

A key aspect of A3's success lies in its structured, disciplined approach, which ensures that no part of the problem-solving process is overlooked. It is used to tackle everything from operational inefficiencies to major strategic challenges, and once completed, the A3 document continues to offer value as both a learning resource and an accountability mechanism.

What it Looks Like

Having been a consultant for over 20 years I have seen a myriad of A3 templates. Everywhere you go they are formatted differently. Some hold true to the Toyota single page format while others are multipage monsters that take your breath away. When doing a web search there are hundreds online. When creating one or leveraging one from someone else, remember the **KISS** rule: **K**eep **I**t **S**imple. The more complex and time consuming something is, the harder it is to sustain from a people management prospective.

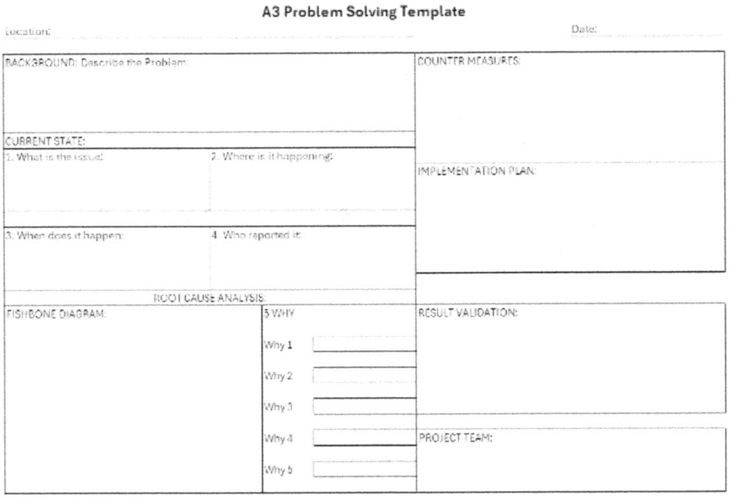

A3 Problem Solving Template

Location: Date:

BACKGROUND: Describe the Problem:

COUNTER MEASURES:

CURRENT STATE:

1. What is the issue:

2. Where is it happening:

IMPLEMENTATION PLAN:

3. When does it happen:

4. Who reported it:

ROOT CAUSE ANALYSIS:

FISHBONE DIAGRAM:

5 WHY

Why 1

Why 2

Why 3

Why 4

Why 5

RESULT VALIDATION:

PROJECT TEAM:

This is a basic template I've used for years. It is basic and easily understood. Choose a layout that works for your needs. You can amend the headings and formatting anyway you like however once you have one that works, stick with it. You want a consistent process to ensure compliance.

Components of the A3 Template

The A3 problem-solving template consists of several core sections, each designed to guide the user through a structured approach to problem-solving. These sections are arranged logically to ensure that teams fully understand the problem before jumping to solutions. Below, we explore each section in detail, discussing its purpose, how to complete it, and how it drives the overall problem-solving process.

Background

The background section sets the stage by providing context for the problem at hand. It explains why the problem needs to be solved, its impact on the organization, and the broader business implications. This section is crucial for aligning everyone on the team and stakeholders with the significance of the issue.

- **Purpose:** The background explains the problem's relevance to the organization. It answers questions like: Why are we addressing this issue? What business processes or outcomes are being affected?
- **Key Elements:**
  - **Problem Identification:** Define the issue in a clear and concise manner. Focus on facts rather than symptoms.
  - **Business Impact:** Highlight how the problem affects key business metrics such as cost, quality, delivery times, or customer satisfaction. This makes the case for why resolving it is critical.
  - **Context:** Provide relevant historical or situational context, especially if this problem has occurred before or if it ties into larger strategic goals.

How to Use It:

In practice, this section is used to engage stakeholders and decision-makers, showing them why the problem is important enough to warrant resources and attention. Be sure to include measurable impacts (e.g., "Delays in the production line have increased lead times by 20%, resulting in missed customer deadlines"). This encourages stakeholders to take the problem seriously and ensures alignment on the need for resolution.

Current State

The current state section captures the existing situation with as much detail as possible, providing a baseline for comparison once solutions are implemented. This is where the problem is thoroughly analyzed and visualized.

- **Purpose:** The goal of the current state is to provide a comprehensive snapshot of how things currently stand. This includes both qualitative descriptions and quantitative data.

- Key Elements:
    - **Process Mapping:** Use tools like flowcharts or process maps to illustrate the existing workflow or operations. Visuals are particularly useful in this section to make complex processes easier to understand.
    - **Data Collection:** Gather data that illustrates the extent of the problem. This could include error rates, downtime, quality control failures, or other key performance indicators (KPIs) relevant to the issue.
    - **Observations:** Include insights from those closest to the problem. Interviews with employees, operators, or team members involved can provide qualitative data that numbers alone might not reveal.

How to Use It:

In practical application, this section is crucial for identifying where bottlenecks, inefficiencies, or gaps exist. For example, if the problem is related to equipment downtime, this section would outline how often the breakdowns occur, how long they last, and what operational impact they have. By using data to support the observations, it becomes easier to pinpoint where corrective actions are needed.

Goal (Target State)

The goal section defines the desired outcome or target state once the problem is solved. It sets specific objectives that the solution must achieve, ensuring that everyone involved understands what success looks like.

- **Purpose:** To clearly define what the team aims to accomplish. Without a well-defined goal, it is difficult to determine whether the solutions implemented have been effective.
- **Key Elements:**
    - **SMART Goals:** Objectives should be Specific, Measurable, Achievable, Relevant, and Time-bound. For example, "Reduce lead times by 15% within the next three months."
    - **Metrics for Success:** Outline the key performance indicators (KPIs) that will be used to measure whether the solution has been successful. Ensure that these metrics are related to the problem identified in the current state.
    - **Alignment with Business Goals:** The goals set here should align with broader organizational objectives. Whether it is improving efficiency, reducing costs, or enhancing customer satisfaction, the goal should tie into the larger strategic picture.

How to Use It:

This section is where you align the team on a clear outcome. For instance, in a manufacturing setting, a goal might be to "Reduce defect rates from 8% to 2% over the next quarter." This sets a clear target that will drive the actions taken later in the problem-solving process. Defining goals in this way also makes it easier to measure progress and results once the solution is implemented.

Root Cause Analysis

Root Cause Analysis (RCA) is the heart of the A3 template. In this section, teams dive deep into the underlying causes of the problem rather than merely addressing its symptoms. Common tools used in RCA include the 5 Whys and the Fishbone Diagram.

- **Purpose:** The RCA section seeks to answer the question: "What is causing this problem?" The goal is to uncover the root cause(s) to ensure that the proposed solutions address the core issue and not just surface-level symptoms.
- **Key Elements:**
  - **5 Whys Method:** This method involves asking "Why?" repeatedly to trace the problem back to its root cause. For example, if the problem is frequent machine breakdowns, asking "Why did the machine break down?" repeatedly might reveal that inadequate maintenance schedules are the root cause.
  - **Fishbone Diagram (Ishikawa Diagram):** This visual tool helps teams categorize plausible causes under broad headings such as "People," "Processes," "Equipment," and "Materials." It is particularly useful for complex problems with multiple contributing factors.

How to Use It:

The RCA section requires input from cross-functional teams, as different perspectives can shed light on various aspects of the problem. For example, if a bottleneck in production is the issue, involving maintenance, quality control, and operations teams can help reveal whether the problem is due to equipment failure, lack of operator training, or supply chain disruptions. By drilling down into the root causes, teams can avoid implementing "band-aid" solutions and focus on fixing the underlying problem.

Countermeasures

Countermeasures are the solutions or actions proposed to address the identified root causes. Rather than listing general ideas, this section requires specific, actionable steps that will eliminate or reduce the impact of the problem.

- **Purpose:** The countermeasures section is where teams propose and document actionable solutions to the problem. Each solution must be directly linked to a root cause identified in the previous section.
- **Key Elements:**
    - **Actionable Steps:** Propose practical solutions that can be implemented within the organization's current capabilities and resources. For example, if poor operator training was identified as a root cause, a countermeasure might be to implement a new training program.
    - **Feasibility and Impact:** Assess each countermeasure for its feasibility and potential impact. Solutions that are too costly or difficult to implement may need to be reconsidered.
    - **Link to Root Causes:** Ensure that each countermeasure is directly connected to the root cause(s) identified. Avoid generic solutions that do not address the specific issue.

How to Use It:

In practice, this section helps ensure that the solutions are both feasible and effective. For example, if one of the root causes is equipment failure due to poor maintenance, a countermeasure might be to implement a preventive maintenance schedule with clear guidelines and accountability. This way, the solution addresses the root cause rather than just treating the symptoms.

Implementation Plan

This section details how the proposed countermeasures will be executed. It includes specific action items, responsible parties, timelines, and resources needed for successful implementation.

- **Purpose:** The implementation plan outlines how the countermeasures will be put into action. It ensures that the solutions are executed properly and on time.
- **Key Elements:**
  - **Action Items:** Break the solution down into specific steps or tasks that need to be completed. Assign responsibility for each task to specific individuals or teams.
  - **Timelines:** Establish a clear timeline for each task, including start and end dates. This ensures that the team stays on track.
  - **Resources Required:** Identify any resources (e.g., budget, personnel, equipment) that are needed to carry out the countermeasures.

How to Use It:

This section ensures that the solution does not just remain a proposal, it gets implemented. For example, if the solution involves upgrading equipment, the implementation plan might include action items like "Research new equipment options," "Obtain approval for budget allocation," and "Train operators on new machinery." Each task is assigned to a specific person or team, with deadlines for completion.

Effect Confirmation

Effect confirmation is where the results of the implemented countermeasures are evaluated. This section compares the current state after implementation to the original goals and determines whether the solutions were successful.

- **Purpose:** To evaluate whether the implemented countermeasures achieved the desired results and resolved the problem. This section answers the question: "Did the solution work?"

- Key Elements:
  - **Data Comparison:** Compare the original data (before the solution was implemented) with the data collected after implementation. This provides a clear picture of whether the problem has been resolved and by how much. For example, if the goal were to reduce machine downtime by 10%, the data should now show that downtime has decreased by at least that amount.
  - **Deviation Analysis:** If the results did not meet expectations, analyze why the solution fell short. Was there a flaw in the implementation, or were the wrong root causes identified? This step is crucial for adjusting strategies and making improvements.
  - **Sustainability Check:** Determine whether the solution can be sustained over time. Temporary fixes might work in the short term but fail to hold up in the long run, which is why long-term effectiveness must be evaluated.

How to Use It:

This section is essential for measuring success. After solutions are implemented, the team should continue gathering data to assess whether the changes have led to the expected improvements. For example, if a solution involved preventive maintenance schedules, the team would review maintenance records and production data to confirm that machine downtime has decreased as planned. If the solution fell short, the team must revisit the root causes or implementation plan, making necessary adjustments to achieve the goal.

Follow-Up Actions and Sustainability

Once the effectiveness of the solution is confirmed, the next step is to ensure that improvements are sustained over time. This section outlines how the solution will be monitored and what steps will be taken to prevent the problem from reoccurring.

- **Purpose:** To ensure that the corrective actions taken are not just short-term fixes but are embedded into the organization's processes and culture for long-term sustainability.
- **Key Elements:**
    - **Monitoring Plan:** Define how the solution will be monitored moving forward. This could involve setting up regular check-ins, reviews, or audits to ensure that the solution remains effective. Monitoring might also involve tracking key performance indicators (KPIs) over time to ensure continued improvement.
    - **Standardization:** Where applicable, update standard operating procedures (SOPs) or training materials to reflect the new process. This helps institutionalize the solution, ensuring that it becomes part of the organization's way of operating.
    - **Additional Training or Resources:** If the solution requires new skills or knowledge, ensure that employees are adequately trained. For example, if the root cause was related to operator error, the follow-up plan should include continued training and support for those workers.

How to Use It:

In this section, you ensure that the problem does not reoccur. For instance, if the solution involved implementing a new process, the organization would monitor that process regularly to ensure compliance. Additionally, updating SOPs and training employees ensures that everyone understands and follows the new protocols. This approach prevents a regression to the old way of doing things, securing long-term success.

Conclusion

The A3 problem-solving template is a powerful tool for systematically addressing and solving complex problems. Each section—background, current state, goal, root cause analysis, countermeasures, implementation plan, effect confirmation, and follow-up—serves a specific purpose in guiding teams through the problem-solving process. The A3 format encourages clarity, accountability, and continuous improvement, making it an essential tool for business leaders looking to foster a culture of operational excellence.

By mastering each section of the A3 template, teams can develop comprehensive solutions that not only address the root causes of problems but also ensure that improvements are sustained over time. This holistic approach enables organizations to solve current issues while building a foundation for long-term success.

Chapter 5: Right Mindset

Introduction

The mindset you bring into a root cause analysis (RCA) session is critical to its success. RCA is not just about finding what went wrong—it is about creating an environment where every participant feels safe to share their insights without fear of judgment. Without this, the genuine issues remain buried beneath defensiveness and blame, and the opportunity to resolve them may be lost. Equally important is fostering a sense of urgency and ownership in the room. The success of RCA depends on not only identifying the problem but on every person in the room taking responsibility for driving the solution forward.

Creating the Right Environment

The key to a successful RCA session is psychological safety. This means creating a space where people feel confident enough to speak openly, without worrying about being blamed or judged. When participants feel safe, they are far more likely to offer the candid insights necessary to uncover the root cause of an issue. The facilitator plays a pivotal role in this. By setting the right tone from the start, you ensure the team feels comfortable, and this is where you begin laying the groundwork for solving the problem.

Blame has no place in an RCA session. If people feel they are being singled out or criticized, they will retreat, and you will end up with surface-level insights, rather than the deeper truths that are critical to fixing the underlying issue. For the session to work, everyone in the room must know that the focus is on finding solutions, not assigning fault. Facilitators should make this clear right from the outset.

The Facilitator's Role: Encouraging Ownership and Urgency

The leader's responsibility extends beyond simply guiding the conversation. They are there to ensure that the people in the room know they own the outcome of the session. This ownership starts with the mindset that solving the problem is not just an intellectual exercise, it is a necessity for the organization to improve and move forward.

An effective facilitator helps the team understand that while they are here to solve the immediate problem, the insights they uncover and the actions they take have long-term impacts. Facilitating with a sense of urgency communicates that every person in the room has a direct role in preventing future issues by addressing this one thoroughly and honestly today.

Equally important is the understanding that the solution does not rest with the facilitator alone. While you guide the session, it is the participants who will own the solutions, and they must leave the room knowing that their input directly influences the outcome. This is why you must encourage open participation from every member—getting everyone engaged and contributing is non-negotiable. People must understand that their voices are crucial to solving the problem, and that the quality of the solution directly depends on the quality of their input.

Urgency Through Action Plans and Accountability

An RCA session is not just a discussion—it is the starting point for action. Once the root cause has been identified, it is critical that the conversation translates into clear, actionable steps. Every action item must have a designated owner, a specific deliverable, and a timeline. Without this, the session loses its impact, and the problems remain unresolved.

To ensure follow-through, accountability must be emphasized. This is where the urgency comes into play. Participants need to leave the room knowing that the actions identified are not "suggestions," they are non-negotiable tasks that need to be completed. I always close the session by reviewing the action plan with the group, assigning ownership, and setting deadlines. It is essential to schedule regular check in meetings after the session to track progress, to ensure actions are on schedule, actions are completed, the results meet expectations and any roadblocks are removed.

Summary

The mindset you establish going into an RCA session will determine its success. By fostering an environment of psychological safety, you create the conditions for open dialogue, ensuring that participants feel free to contribute without fear of blame. At the same time, you must deliver a clear sense of urgency and ownership, helping the team understand that their input is critical to solving the problem and preventing its recurrence.

The facilitator guides the conversation and set the stage for meaningful participation. Through prewritten flipchart pages, you visually anchor the team to the goals, ground rules, and expectations for the session. Action plans, accountability, and a sense of urgency drive the team toward results, and tools like the Parking Lot help manage scope while maintaining momentum. Ultimately, the session's success depends on your ability to create a space where everyone is engaged, responsible, and ready to contribute to lasting solutions.

Chapter 6: Defining the Problem Statement

Introduction

When problems arise in business, the instinct is to jump in and solve them quickly. It is a natural reaction: we are conditioned to think that the faster we act, the sooner things will improve. But this rush to fix things often comes at a cost. The real danger is not the problem itself—it is misunderstanding the problem. How many times have you seen teams rush to fix something, only to find the issue reappears, sometimes worse than before?

The reason is simple: you cannot fix what you do not fully understand. Defining the problem statement is not just an early step in the problem-solving process; it is the most critical one. Without a clear understanding of what the real issue is, even the best-designed solutions will fail to address the root cause.

This chapter will explore why defining the problem statement is not just a procedural step but a strategic necessity. We will talk about the common pitfalls of rushed problem-solving, how to break down surface-level issues, and offer guidance on how to define a problem statement that sets the stage for successful, lasting solutions.

The Danger of Fixing the Wrong Problem

In the rush to solve problems, many leaders make the mistake of attacking symptoms rather than root causes. They see an issue and immediately jump to fix it without stepping back to ask, "Is this really the problem?" This leads to quick, short-term fixes that do not stand the test of time. Worse yet, it can exacerbate the underlying problem, creating a vicious cycle of recurring issues.

What is at Stake?

When the wrong problem is addressed, it drains resources—time, money, and energy. You may find that after implementing a solution, the problem persists, leading to frustration for both your team and stakeholders. Worse, solving the wrong problem can damage customer relationships, erode employee morale, and create confusion across departments.

A Case in Point:

Imagine a software company noticing a spike in customer complaints about product usability. The immediate reaction might be to overhaul the product's interface. Teams rush to make the product more user-friendly, only to discover after months of work that the real issue was not the product at all—it was lack of proper onboarding and training for customers. By misdiagnosing the issue, the company wasted resources on a solution that did not address the underlying problem.

The lesson here is that problem-solving without a well-defined problem is like driving with a blindfold. No matter how fast you go, you will never reach your destination.

Understanding What the Problem Really Is

A clearly defined problem statement serves as the foundation for any successful solution. It is not enough to describe what is happening on the surface; the key is to understand what lies beneath.

Breaking Down Surface-Level Symptoms vs. Root Problem

Surface-level symptoms are the things we notice first—the obvious signs that something is not right. But these are rarely the actual problem. A strong problem statement digs deeper, asking the critical question: "What is really going on here?"

Let us revisit the earlier example of the retail company with a spike in returns. On the surface, the issue might appear to be about product quality. You may believe customers are returning items because the products are not meeting expectations, however a deeper investigation might reveal something else.

Example (Expanded):

Let us say this retail company has seen a 25% increase in customer returns over the last six months. Initially, leadership assumes the problem is poor product quality. They spend time and money improving the materials and manufacturing processes. Yet, returns continue to rise. Upon further examination, the real issue turns out to be inaccurate product descriptions on the company's website. Customers were misled by the product details, so the returns were a result of mismatched expectations rather than defective products.

In this case, the real problem was not product quality—it was a communication breakdown between the marketing team and the product developers. Defining the problem accurately would have led to a solution focused on improving how products were described online, rather than making unnecessary changes to manufacturing.

The key takeaway? The obvious issue is rarely the actual issue. If you want a lasting solution, you need to take the time to uncover the true problem.

Defining the Problem Statement: A Strategic Process

Defining the problem statement is not just about identifying what is wrong—it is about crafting a clear, actionable statement that aligns everyone on the team toward solving the right issue. A well-defined problem statement answers three critical questions:

1. What is the Problem?
2. Why is it a problem?
3. What will happen if we do not address it?

The Components of a Strong Problem Statement

- **What (The Issue Itself):** Start by clearly stating what is going wrong. Avoid jargon or assumptions. The goal is to describe the problem in its simplest, most factual form. For example, "Sales have decreased by 15% over the past quarter," or "Production output has dropped by 20%."

- **Why (The Impact on Business):** Explain why this problem matters. What is the consequence of this issue continuing? Does it affect revenue, customer satisfaction, employee efficiency, or product quality? Providing context is key to helping everyone understand the importance of solving the problem. For example, "The decrease in sales is leading to a loss of market share and decreased profitability."
- **Consequence of Inaction:** Articulate what will happen if the problem goes unsolved. This adds urgency to the problem-solving process, helping to prioritize resources and focus efforts. For example, "If production output continues to drop, we risk missing delivery deadlines, which could lead to customer attrition."

Crafting the Problem Statement

A problem statement should be concise, clear, and focused on facts. It should avoid attributing blame or jumping to conclusions about the cause. A good format to follow is:

- **The Situation:** We are currently experiencing [issue] with [impact].
- **The Impact:** This problem is affecting [business function] by [consequence].
- **The Stakes:** If we do not solve this, we risk [long-term consequence].

For example:
"We are currently experiencing a 15% drop in sales over the past three months, which is affecting our revenue and market share. If we don't address this, we risk further financial losses and a weakened competitive position in the market."

This format provides clarity on the problem without making assumptions about the root cause. It allows teams to dig deeper and investigate why the issue is happening, rather than rushing into solutions.

Common Pitfalls When Defining the Problem Statement

Even when leaders understand the importance of a clear problem statement, there are several common pitfalls to watch out for:

- **Assumptions Without Evidence:** Many problem statements fail because they include assumptions about the cause. For instance, stating, "The problem is that our employees aren't working hard enough," is an assumption. A better statement would be, "Our productivity has decreased by 10%, and we need to understand why."
- **Focusing on Symptoms:** As discussed earlier, it is easy to focus on the most obvious aspects of a problem. The challenge is to look beyond symptoms and define the underlying issue. For example, do not just say, "Our customers are returning products"—ask why.
- **Overcomplicating the Statement:** A problem statement should be simple and straightforward. If it is too long or filled with technical jargon, it can confuse the team and lead to misalignment. Keep it factual and avoid unnecessary details.

Building a Problem-First Culture

Defining a clear problem statement is not just an isolated task—it is part of building a problem-first culture within an organization. When teams are encouraged to pause and reflect before jumping into solutions, they are more likely to find long-term success.

Encouraging Teams to Ask Questions

Leaders should foster an environment where asking questions is valued more than offering quick fixes. Questions like "What are we really solving here?" or "What's the consequence of doing nothing?" can shift the focus from reactive to initiative-taking problem-solving.

Aligning the Organization:

A strong problem statement helps align cross-functional teams by providing a shared understanding of the issue. When everyone from sales to product development understands what is at stake, they are more likely to collaborate effectively toward a common goal.

Clarity as the Cornerstone of Success

Defining the problem statement is not just a step in the Root Cause Analysis process—it is the foundation upon which successful solutions are built. Without clarity, organizations risk addressing symptoms rather than root causes, leading to wasted resources, missed opportunities, and recurring issues. The problem statement acts as a compass, ensuring that teams focus their energy and resources on solving the right issue.

Crafting a strong problem statement demands patience, collaboration, and a willingness to question assumptions. It is about digging deeper to uncover the "why" behind the "what," moving beyond surface-level symptoms to get to the heart of the matter. This clarity not only aligns teams but also fosters a culture of thoughtful, strategic problem solving where challenges are met with curiosity and solutions are designed to last.

The power of a well-defined problem statement lies in its ability to unite stakeholders, direct efforts, and inspire action. By embracing the principles outlined in this chapter, business leaders can transform their approach to problem-solving, ensuring that each challenge becomes an opportunity for meaningful improvement. Remember, success begins not with the solution, but with a clear understanding of the problem.

Chapter 7: Who Should Attend the RCA Session and Why

Introduction

The success of any Root Cause Analysis (RCA) session depends not only on the tools and processes but also on the people involved. Choosing the right participants is vital for uncovering root causes, fostering collaboration, and crafting practical solutions. Without the right mix of expertise, perspectives, and authority, even the best-facilitated session can fall short of its potential.

Core Criteria for Attendance

The success of a Root Cause Analysis (RCA) session is influenced by the people in the room. The right mix of participants brings the depth and breadth of expertise needed to uncover and address root causes effectively. Selecting attendees is not about filling seats—it is about curating a group that will provide valuable insights, contribute to meaningful discussions, and drive actionable outcomes.

1. **Subject Matter Experts (SMEs):**
   SMEs are the backbone of an effective RCA session. These are the individuals who possess in-depth knowledge about the processes, systems, or operations affected. For example, if the issue relates to equipment failures, bringing in a maintenance specialist is crucial. Their firsthand experience can shed light on nuances that others might overlook, such as subtle patterns of wear or operational quirks. SMEs ensure that the discussion remains grounded, avoiding overly theoretical or impractical solutions.

2. **Cross-Functional Representation:**
   Problems rarely exist in silos, and solutions should not be developed in silos either. Including representatives from different departments or functions ensures that the analysis considers all angles. For instance, a manufacturing defect might stem from a combination of supply chain inconsistencies, operational inefficiencies, and quality control lapses. Having voices from procurement, production, and quality assurance in the room ensures a holistic understanding of the issue and promotes solutions that work across boundaries.

3. **Decision-Makers and Stakeholders:**
   To move from analysis to action, you need participants who can make decisions and allocate resources. Including managers or team leads ensures that any proposed solutions have the buy-in needed for swift implementation. Decision-makers can also provide strategic context, aligning RCA outcomes with broader organizational objectives. Without their presence, even the most effective solutions may struggle to gain traction.
4. **Frontline Employees:**
   Often, the most valuable insights come from those who are closest to the problem. Frontline employees—whether they are operators, customer service representatives, or technicians—bring a practical perspective. They understand how things work on the ground and can highlight issues or workarounds that others might not even be aware of. Their inclusion also fosters a sense of ownership, increasing the likelihood of successful implementation.

5. **Facilitator:**
   While not directly contributing to the root cause analysis, the facilitator plays a critical role in guiding the session, maintaining focus, and ensuring a productive discussion. A skilled facilitator brings neutrality, encourages balanced participation, and keeps the group aligned with the session's objectives. Without effective facilitation, even the best-selected participants can become derailed by groupthink, dominant voices, or tangential debates.
6. **Optional but Valuable Attendees:**
   In some cases, it may be helpful to include external perspectives, such as consultants, industry peers, or even key customers. These individuals can provide an unbiased view or highlight industry benchmarks that challenge internal assumptions. However, their participation should be carefully evaluated.

Why It is Important to Include Outside of the Leadership Team

A successful RCA session thrives on the collective knowledge of its participants. By including individuals from a variety of roles, departments, and levels within the organization, you gain a broader perspective on the issue at hand. This diversity of viewpoints allows the group to see the problem from multiple angles and uncover insights that might otherwise remain hidden.

Dissimilar roles often interact with the same problem in unique ways. For example, frontline employees may see recurring operational issues that managers are not aware of, while leadership might provide insights into strategic pressures affecting decision-making. A maintenance worker could identify a subtle pattern in equipment breakdowns that aligns with scheduling inconsistencies highlighted by a planner. Bringing these perspectives together not only enriches the analysis but also ensures that critical details are not overlooked.

Beyond problem-solving, having a broad mix of attendees fosters collaboration across the organization. When individuals from different departments or hierarchy levels come together, it builds mutual understanding and respect. This shared experience of working toward a common goal can break down silos and establish a sense of unity. Everyone at the table becomes a valued contributor to the solution, which increases buy-in and accountability for implementing the agreed-upon actions.

Moreover, inviting participants from different levels within the organization sends a powerful message: everyone's input matters. This inclusivity builds trust and shows that the organization values the contributions of all its members, regardless of their role. It also helps create champions for the changes that will result from the session, as people are more likely to advocate for solutions they helped develop.

In essence, the strength of an RCA session lies not just in the skills or expertise of its attendees but in the diverse perspectives they bring and the collaborative culture it fosters. By uniting individuals from across the organization, you not only solve the immediate problem but also strengthen the team's capacity for tackling future challenges together.

Chapter 8: Facilitating the RCA Session

Introduction

Root Cause Analysis (RCA) sessions are where problems are dissected, perspectives are shared, and actionable insights emerge. The success of these sessions hinges not just on the tools or methods employed, but on how effectively the discussions are facilitated. The facilitator is the linchpin of the process—the person who ensures the session stays focused, productive, and inclusive, while guiding participants toward uncovering the true root cause of the problem.

Facilitating an RCA session is not just about managing a room; it is about creating an environment where everyone feels heard and valued. It requires balancing objectivity with empathy, fostering open communication while maintaining structure, and ensuring the team remains solution-oriented rather than descending into blame. A skilled facilitator understands that the purpose of RCA is not only to find answers but also to build trust and collaboration among stakeholders.

In this chapter, we will explore the critical role of the facilitator in leading a successful RCA session. From setting the stage and managing group dynamics to leveraging tools like the Fishbone Diagram and 5 Whys, you will learn how to guide your team through the RCA process with confidence and clarity. Whether you are stepping into this role for the first time or refining your approach, this chapter will equip you with the strategies and techniques to make every RCA session a constructive and impactful experience.

Setting the Stage

The success of a Root Cause Analysis (RCA) session begins long before the participants gather in the room. A well-prepared facilitator sets the stage by creating a conducive environment, ensuring participants feel comfortable and engaged, and laying the groundwork for productive discussions. Whether it is choosing the right time and place, fostering trust through ground rules, or having the necessary materials on hand, these elements play a pivotal role in driving the session's success.

The Physical Environment: Creating the Right Space

The space where the RCA session takes place should be more than just functional—it should encourage collaboration and creativity. Select a room that is well-lit, spacious enough to accommodate all participants comfortably, and free from distractions. Arrange seating in a way that fosters open communication, such as a U-shape or circular format, ensuring everyone feels included and equal.

Think about amenities that make the session comfortable: access to water, coffee, or light snacks can help participants stay focused. A whiteboard or flipcharts should be prominently displayed, along with any visual aids or templates you will use during the session.

Timing: Picking the Optimal Moment

Timing is crucial for an effective RCA session. Schedule the session at a time when key participants are available and can fully engage without pressing distractions. Avoid scheduling the session during particularly hectic times in the organization, as competing priorities can reduce focus and participation. Sessions held earlier in the day often benefit from participants being fresher and more energetic.

It is also important to set realistic time expectations. RCA sessions can run longer than anticipated, so plan for breaks and ensure there is enough time to explore all aspects of the issue without rushing.

Ground Rules: Building Trust and Ensuring Productivity

Establishing ground rules at the outset of the session fosters trust and creates an atmosphere where participants feel safe to speak openly. Here are some core principles to consider:

- **No Blame Culture**: Make it clear that the purpose of the session is to uncover systemic issues, not to assign fault to individuals. Use language like, "We're here to find out *what* happened, not *who* caused it."
- **Respectful Communication**: Encourage participants to listen actively and respect diverse perspectives. Remind them that differing opinions can lead to better outcomes.
- **Focus on Facts**: Emphasize that discussions should be evidence-based, steering away from speculation or assumptions.
- **Participation is Key**: Encourage all voices to be heard. Ensure that no single person dominates the conversation and that quieter team members are invited to share their insights.
- **Limit Distractions:** I also ask participants to turn off phones and computers unlet they are required for photos and relevant data of the issue.

Reinforce these ground rules throughout the session, and if necessary, intervene gently to redirect conversations or mediate disagreements.

Materials: Preparing for a Productive Session

Having the right materials on hand ensures the session runs smoothly and stays focused. As a facilitator, your toolkit might include:

- **Visual Aids**: Flipcharts, whiteboards, markers, and sticky notes are essential for capturing ideas, mapping processes, and visually organizing information.
- **Templates and Checklists**: Bring pre-prepped RCA tools, such as Fishbone Diagrams, the 5 Whys template, or timeline analysis charts. These help structure the discussion and keep it on track.
- **Pre-Prepared Questions**: Develop a list of open-ended, probing questions tailored to the issue at hand. For instance, "What were the conditions leading up to this event?" or "What patterns do we see in the data?"
- **Reference Materials**: Provide participants with data, reports, or other documentation relevant to the issue. This ensures discussions are grounded in facts.

Example: Prepping for a Manufacturing Downtime RCA Session

Imagine facilitating an RCA session to address unexpected equipment downtime in a manufacturing plant. As the facilitator, you would:

1. **Set Up the Room**: Arrange a conference room with a flipchart labeled "Downtime RCA," along with sticky notes and markers for brainstorming.
2. **Share Key Materials**: Provide participants with production logs showing downtime trends, maintenance schedules, and operator reports for context.

3. **Establish Ground Rules**: Begin by emphasizing that the session is about improving processes, not assigning blame, and remind the group to stay focused on facts.
4. **Plan the Agenda**: Allocate time for each step, from identifying symptoms to brainstorming potential causes, ensuring there's time for discussion without veering off track.

By carefully setting the stage, you create an environment where participants feel empowered to collaborate, problem-solve, and uncover the root cause of the issue. When everyone feels engaged and supported, the session becomes a powerful tool for driving meaningful improvements.

Prewritten Flipchart Pages

One practical way to maintain focus and set the right tone is through prewritten flipchart pages. These is a simple but powerful tool to visually communicate expectations and guide the session. Before the RCA session begins, prepare key flipchart pages that help set the stage:

1. **Goals and Objectives**:
   Start by outlining the key goals on a flipchart. This serves as a visual reminder throughout the session of what the team is aiming to accomplish. These goals might include "Identify the root cause of [issue]" or "Develop actionable solutions to prevent recurrence."

2. **Ground Rules**:
   List essential ground rules for the session. For example, "Open and honest conversations", are key. After presenting the rules, I invite participants to add their own, reinforcing that this is a collaborative process. This involvement helps foster a sense of shared responsibility.

3. **Expectations:**
   A flipchart labeled "Expectations" is another useful tool. I include points like "Everyone participates" and "Respect all opinions." This helps set the tone where everyone knows they are expected to contribute. Asking participants to add to this list is a great way to further establish ownership and accountability for the session's outcome.

4. **Action List**
   Prepare a page to capture action items. Refer to chapter 10 on creating action plans.

5. **Parking Lot:**
   During an RCA session, it is common for off topic issues or unrelated concerns to surface. While it is important to stay focused on the primary issue at hand, you do not want to dismiss valuable insights that may emerge. To manage this, use a "Parking Lot" flipchart. When topics come up not part of the active discussion, note them for future discussion. This ensures they are acknowledged without disrupting the session's flow.

At the end of the meeting, review the Parking Lot items and either address them if time permits, or add them to the action log for follow-up. This technique maintains focus while showing participants that their concerns are being heard and will be addressed at the right time.

These prewritten flipcharts serve as constant reference points during the meeting. They keep participants focused, reinforce the need for open dialogue, and provide visual cues that ground rules and expectations are more than just words—they are the framework for success in the session.

Facilitator's Role in an RCA Session

The facilitator plays the vital role in guiding the Root Cause Analysis (RCA) session to success. The facilitator is responsible for maintaining neutrality, encouraging balanced participation, and managing group dynamics to ensure that the session remains productive and focused on uncovering the root cause. This role requires both interpersonal and organizational skills, as the facilitator must navigate diverse perspectives while steering the conversation toward actionable outcomes.

Maintaining Neutrality and Guiding the Discussion

At the heart of effective facilitation lies neutrality. As a facilitator, your role is not to impose opinions or sway the group toward a particular outcome, but to ensure that the discussion remains objective and focused on the issue at hand. Neutrality fosters a sense of trust among participants, creating an environment where everyone feels their contributions are valued, and where candid discussion can flourish.

One way to maintain neutrality is by steering the conversation toward understanding the process rather than attributing blame. For example, instead of letting the discussion drift into "Who caused this problem?" redirect with "What factors contributed to this situation?" This subtle shift reframes the dialogue to focus on systemic issues rather than individual accountability, keeping the atmosphere collaborative rather than confrontational.

Facilitating a structured conversation is equally critical. RCA sessions can cover a wide range of topics, so breaking the discussion into distinct phases—such as identifying symptoms, analyzing causes, and brainstorming solutions—helps maintain clarity and purpose. Without structure, it is easy for participants to jump to conclusions or digress into tangential topics. A facilitator's role is to gently guide the group back on track when this happens, ensuring that the session progresses logically and efficiently.

The art of asking the right questions is another cornerstone of effective facilitation. Open-ended questions like, "What trends or patterns are emerging from the data?" or "What might we be overlooking?" encourage deeper exploration and critical thinking. Such questions not only stimulate discussion but also uncover perspectives or details that may not have been immediately apparent.

For instance, in an RCA session focused on reducing customer complaints, if the group fixates on a single issue like delivery delays, a facilitator might interject with, "That is an important factor. What about other touchpoints in the customer journey? Could there be additional contributors?" This approach broadens the scope of the discussion and ensures a more comprehensive understanding of the problem.

Ultimately, maintaining neutrality and guiding the discussion are about creating a space where ideas can be exchanged freely, without fear of judgment or dismissal. It is about ensuring that every voice is heard, every angle is considered, and the group remains united in their goal of uncovering the root cause and devising actionable solutions. A skilled facilitator transforms a room full of diverse stakeholders into a cohesive team, aligned in purpose and driven to make meaningful improvements.

Balancing Participation from All Stakeholders

One of the most challenging aspects of facilitation is ensuring that all stakeholders are equally engaged. Diverse perspectives are essential for uncovering root causes, so the facilitator must create opportunities for everyone to contribute.

- **Encouraging Quiet Voices**: Some participants may hesitate to share their insights, especially in a group setting. The facilitator can gently invite them to speak by saying, "I'd like to hear your thoughts on this," or "What's your perspective on this aspect?"
- **Preventing Domination**: Strong personalities can sometimes dominate the conversation, unintentionally silencing others. Intervene tactfully by acknowledging their contribution and opening the floor: "That's an excellent point—let's hear from others to build on that."
- **Using Structured Tools**: Incorporating tools like round-robin discussions or written brainstorming exercises ensures equal participation. For example, participants could write down their thoughts on sticky notes before sharing with the group.

Example: In a healthcare RCA session, a senior surgeon might dominate the discussion on surgical errors, overshadowing the insights of nurses or technicians. The facilitator could balance participation by introducing a roundtable approach, where each stakeholder shares their observations and experiences.

Managing Group Dynamics and Resolving Conflicts

RCA sessions often bring together stakeholders with varying priorities, perspectives, and emotional investments. Managing these dynamics effectively is critical to maintaining a constructive atmosphere and driving the session toward actionable outcomes.

- **Addressing Tensions Constructively**: Disagreements are natural but can derail the session if not managed carefully. Acknowledge differing viewpoints by saying, "It's great that we're exploring this from multiple angles—let's delve into the data to see what aligns with the facts."
- **Reinforcing Ground Rules**: If discussions become heated or unproductive, remind the group of the ground rules established at the start of the session. For example, "Remember, we agreed to focus on systemic issues rather than individual blame."
- **Identifying Common Goals**: When conflicts arise, refocus the group on shared objectives. Reiterate that the goal is to improve processes and outcomes for everyone involved, rather than assigning fault or proving a point.

Example: During an RCA session on production defects, the quality control team and operations team may clash over responsibility. The facilitator could defuse the tension by saying, "Both of your perspectives are valuable. Let us map out the entire process to see where improvements can be made collectively."

The Facilitator as a Catalyst for Success

The facilitator's role extends beyond simply managing the session; they serve as a catalyst for collaboration, creativity, and problem-solving. By maintaining neutrality, encouraging balanced participation, and skillfully managing group dynamics, the facilitator ensures that the RCA session remains focused on identifying the root cause and developing actionable solutions. A well-facilitated session not only uncovers the underlying issues but also strengthens trust and collaboration among participants, laying the groundwork for effective implementation and long-term success.

Structuring the RCA Session

A well-structured RCA session is key to maintaining focus, maximizing productivity, and ensuring actionable outcomes. The session can be divided into three distinct phases—each with its specific objectives and activities—to guide the team through the problem-solving process efficiently. These phases provide a clear framework for participants, fostering a shared understanding and paving the way for meaningful progress.

Beginning: Defining the Problem Statement

The session should begin by grounding everyone in a shared understanding of the problem at hand. As the facilitator, your primary role in this phase is to ensure that the problem statement is clear, concise, and free of assumptions. Start by presenting the problem as previously defined during the preparation stage but allow room for clarifications or refinements based on input from the participants.

For example, if the problem is "inconsistent delivery times," you might ask participants, "Does this accurately capture the issue we've experienced? Is there anything missing?" This step not only ensures alignment but also sets a collaborative tone for the session. Emphasize that the goal is to understand the problem thoroughly before jumping to solutions.

The facilitator must actively steer the group away from assigning blame at this stage. Remind participants that the focus is on the "what" and "why," not the "who," and that the purpose is to improve the system rather than scrutinize individuals.

Middle: Brainstorming and Root Cause Identification

The middle phase of the RCA session is where the team de ves into uncovering the potential factors contributing to the problem. This stage is about understanding—not solving—the issue. As the facilitator, it is crucial to set the tone with a clear message: the goal is to identify the potential causes of the problem, not to jump to solutions. By focusing solely on root cause identification, the team can approach the problem with an open mind, unclouded by preconceived notions.

Begin this phase with a brief reminder: "Today, we are here to understand the 'why' behind this problem. Let us focus on identifying opportunities and root causes. The solutions will come later."

This framing helps prevent participants from prematurely proposing fixes and encourages them to think critically about the underlying issues. Reinforce that the success of the session depends on fully understanding the problem before moving forward. A strong diagnosis will naturally lead to effective solutions.

Techniques for Identifying Causes

While tools like the Fishbone Diagram and the 5 Whys are valuable here, keep their use simple and focused since they are covered in detail elsewhere in this book. For this session, emphasize their purpose to generate a robust list of potential causes.

Start by encouraging brainstorming, asking participants to think broadly about what might contribute to the problem. For example:

- "What factors could be influencing this issue?"
- "Where in the process might there be breakdowns?"

Use the Fishbone Diagram as a guide to organize these ideas into categories such as People, Processes, Equipment, Materials, or Environment. This categorization helps ensure the discussion covers a wide range of potential factors and avoids tunnel vision.

Encourage the team to move systematically through each category, asking probing questions to elicit diverse perspectives. For example, if discussing "Processes," you might ask:

- "Are there any steps in this process that often get skipped or delayed?"
- "What external factors could disrupt this process?"

Building a Comprehensive List

The facilitator's role here is to ensure that the group generates a substantial list of potential causes. The list does not need to be perfect or exhaustive at this point—it needs to reflect the collective knowledge and experience of the group.

Throughout this phase, keep redirecting the discussion back to the root cause exploration if participants begin drifting into solutioning. It is natural for people to want to fix the problem immediately, but rushing to solutions can stifle creativity and limit the effectiveness of the analysis.

For example, if someone suggests, "We should buy better equipment," gently steer the conversation back: "That could be a great solution to explore later. For now, let us focus on what about the current equipment might be causing the issue."

Avoiding Preconceived Notions

A common pitfall in this phase is allowing preconceived notions to dominate the discussion. Team members may enter the session with firm ideas about what is causing the problem, often based on their own experiences or biases. As the facilitator, challenge these assumptions without dismissing them outright.

For instance, if a participant insists, "It's always been a training issue," respond with questions like:

- "What specific gaps in training might be contributing?"
- "Are there other factors we should also consider alongside training?"

This approach validates their input while keeping the discussion open to other possibilities. The goal is to create an environment where every perspective is valued, but no single viewpoint dictates the outcome.

Conclusion of the Middle Phase

By the end of this phase, the group should have a comprehensive list of potential causes, organized in a way that facilitates deeper exploration and analysis later. Emphasize that this list is a foundation for moving forward—it represents the collective understanding of the problem and sets the stage for prioritizing and addressing the most impactful root causes.

With a clear understanding of the problem, the team is better positioned to develop targeted and effective solutions in subsequent steps.

End: Documenting Outcomes and Next Steps

The session concludes with a focus on documenting the results and setting a course for action. Summarize the key findings from the root cause analysis, ensuring that all participants agree on the identified causes. This shared understanding is critical for building commitment to the next steps.

Use templates or structured formats to document the outcomes. For example, create a summary that includes:

- The refined problem statement.
- A list of identified root causes.
- Any immediate actions or further investigations required.

Once the findings are documented, shift the conversation to actionable next steps. Who will take ownership of the next phase? What are the immediate priorities? While the RCA session itself is about diagnosing the problem, outlining preliminary actions ensures momentum is maintained.

End the session by thanking participants for their time and contributions. Reinforce the importance of their input in driving meaningful improvements and emphasize how their work in this session ties into the organization's broader goals.

Summary

By following this structured approach—beginning with problem definition, progressing through brainstorming and analysis, and concluding with documentation and planning—you ensure that the RCA session is focused, productive, and purposeful. The structure not only provides clarity to participants but also enhances the overall quality of the outcomes.

Encouraging Participation and Overcoming Challenges

An effective RCA session thrives on collaboration and diverse input. Every team member's perspective is valuable, and the facilitator plays a critical role in ensuring that all voices are heard. However, fostering participation often comes with its own set of challenges—quieter individuals may hesitate to speak up, dominant personalities might overshadow others, and discussions can veer into subjective territory. This section explores strategies to encourage balanced participation and maintain a fact-driven focus.

Drawing Out Quieter Voices

Quiet participants are often the hidden gems of an RCA session, holding insights that could unlock critical understanding. Yet, they may hesitate to share their thoughts—whether due to shyness, fear of judgment, or uncertainty about the process. As a facilitator, your goal is to create an environment where these voices feel valued and encouraged to speak up.

Start by setting a tone of inclusivity from the very beginning. Let the team know that every perspective matters and that this is a safe space for exploration, not judgment. Something as simple as saying, "All ideas are valuable, and we're here to explore without fear of being wrong," can make a significant difference.

When quieter voices remain silent, sometimes a gentle nudge is all they need. For example, you might ask, "Sarah, you've worked closely with this process—what are your thoughts on the issue?" By addressing someone directly but non-confrontationally, you give them an opening to share without pressure. If individuals still seem hesitant, tools like anonymous input—sticky notes, digital polls, or virtual chat features—can help them contribute ideas without feeling exposed.

Acknowledgment also plays a critical role. When a quieter team member finally speaks up, take a moment to affirm their contribution. Saying something like, "That's a great point; let's explore that further," shows appreciation for their input and encourages continued participation.

Managing Dominant Personalities

In every group, there is bound to be someone who commands the conversation, whether due to enthusiasm, confidence, or sheer volume. While their insights are valuable, unchecked dominance can stifle others' contributions and derail the session's objectives. Facilitating a balanced discussion requires tact and finesse.

From the outset, establish clear ground rules about participation. A simple statement like, "Let's ensure everyone has the chance to share their perspective," sets the expectation for equitable dialogue. If dominance becomes an issue during the session, you might redirect the flow by saying, "Thanks for sharing, Alex. Let us pause here to hear what others think about this."

Structured turn-taking can also help manage overbearing personalities. For instance, go around the table and invite each participant to share one thought before opening the floor to broader discussion. This ensures that everyone's voice is heard without singling out the dominant individual.

If dominance persists, a private conversation may be necessary. Approach it with empathy, framing the feedback positively: “Alex, your input has been valuable, and I want to make sure we also hear from others who might have different perspectives. Can we work together on that?”

Bringing It All Together

Facilitating balanced participation and maintaining a focus on facts requires a mix of initiative-taking planning, real-time adjustments, and sensitivity to group dynamics. When quieter voices are drawn out, dominant personalities are managed tactfully, and discussions remain grounded in evidence, the RCA session becomes a more inclusive, effective, and purposeful experience. By fostering these conditions, you ensure that the team uncovers not just a single perspective, but a comprehensive understanding of the problem.

Documenting the RCA Session

The final phase of an RCA session is documenting the outcomes—a step that solidifies the progress made and ensures everyone is aligned on the path forward. While the energy and engagement of the session itself are crucial, without a clear record of what was discussed, agreed upon, and planned, much of that momentum can be lost. Documentation provides the framework for accountability, follow-through, and sustained improvement.

Capturing Key Takeaways

At the heart of your documentation should be a concise summary of the key takeaways from the session. These takeaways should include the identified root causes, insights uncovered during discussions, and any emerging themes or patterns. This summary serves as the backbone for future actions, ensuring that all participants and stakeholders have a shared understanding of the session’s outcomes.

For instance, if a session identifies that outdated equipment is a significant contributor to production delays, this insight should be clearly recorded. Include any supporting evidence, such as data trends or observations shared by team members, to provide context for the conclusion.

Assigning Tasks and Ownership

An effective RCA session not only identifies root causes but also lays the groundwork for addressing them. As part of your documentation, clearly outline the next steps, including specific tasks that need to be completed. For each task, assign a single owner to ensure accountability. Avoid vague statements like "someone should investigate equipment options." Instead, specify, "John Smith will research three potential equipment upgrades by December 15th."

Assigning ownership is essential for maintaining momentum after the session. It also helps prevent confusion or duplication of efforts, ensuring that every action is purposeful and aligned with the broader goals.

Follow the steps outlined in Chapter 10 on creating an action plan.

Defining Next Steps

Documenting next steps is about more than just assigning tasks—it is about creating a roadmap for follow-up. Include details such as timelines, checkpoints, and criteria for success. For example, if the group decides to pilot a new process, the documentation should specify when the pilot will begin, what metrics will be tracked, and when the results will be reviewed.

Clarity in the next steps ensures that participants leave the session with a clear understanding of their roles and the overall direction of the initiative. It also provides a foundation for future progress reviews, helping to keep the team on track.

Ensuring Alignment and Clarity

After the session, share the documentation with all participants and relevant stakeholders. This can be done via email, a shared project management platform, or during a follow-up meeting. Sharing the documentation promptly helps reinforce the session's outcomes while the discussions are still fresh in everyone's minds.

Encourage participants to review the documentation and provide feedback. Are the key takeaways accurate? Are the assigned tasks and next steps realistic and clearly defined? Addressing these questions ensures that everyone is aligned and that the session's outcomes are actionable.

Leveraging the A3 Template for Documentation

To tie together the insights and actions from the RCA session, leveraging the A3 problem-solving template can serve as an effective summary document. The A3 template is not just a structured way to organize information; it is a living document that encapsulates the problem, root causes, and action plan in a single, accessible format. Using the A3 template ensures consistency, clarity, and a seamless transition from analysis to execution.

Chapter 4 on the A3 template gives clear guidance on how to leverage the template. Here is a high-level overview:

1. **Background**:
   Begin by briefly summarizing the context of the problem as discussed in the session. Highlight the issue, its impact on the organization, and why resolving it is critical. This section helps everyone align on the significance of the problem and sets the stage for the solutions.

2. **Current State**:
   Document the baseline data collected before the session. Include key metrics or observations that define the problem. This serves as the reference point for measuring the impact of the solutions.
3. **Root Cause Analysis**:
   Populate this section with the root causes identified during the session. Use tools like the Fishbone Diagram or the 5 Whys to summarize the findings. Ensure that each root cause ties directly to the identified problem.
4. **Proposed Countermeasures**:
   Summarize the potential solutions discussed in the session. Focus on addressing the root causes without delving into implementation specifics just yet.
5. **Action Plan**:
   Refer to Chapter 10 for detailed steps on creating an action plan. Use this section to define tasks, assign responsibilities, and set timelines, ensuring alignment with the broader objectives outlined in the RCA.
6. **Follow-Up and Monitoring**:
   Include details about how progress will be tracked and success measured, tying back to the KPIs established in Chapter 9 on quantifying potential solutions.

Why the A3 Template Matters

Incorporating the A3 template as the summary document not only organizes the outputs of the RCA session but also creates a standardized way to communicate and execute the findings. By tying it to actionable steps and aligning it with the broader goals of the organization, the template becomes a bridge between the analysis and implementation phases, ensuring no insight is lost and that the team remains focused on achieving meaningful, measurable outcomes.

Keeping Documentation Alive and Visible

The documentation from an RCA session, particularly the A3 template and the action plan, should not be treated as static records that fade into obscurity. Instead, these documents should serve as dynamic tools that actively guide progress, ensure accountability, and maintain transparency. Their value lies not only in the initial creation but also in their ongoing use to drive meaningful and measurable change.

One of the most effective ways to keep RCA documentation relevant is by posting it in highly visible locations. This could be a physical space, such as a team bulletin board, or a virtual platform, like a shared workspace or project management tool. Visibility keeps the focus on the RCA process and ensures that everyone involved—whether they were part of the original session or not—can see how the effort is progressing. When documentation is prominently displayed, it encourages accountability among team members, reminding everyone of their responsibilities and timelines. It also fosters transparency by informing others about the problem, the root causes identified, and the planned solutions, creating an opportunity for shared learning and collaboration.

Incorporating RCA documentation into regular team meetings or progress reviews further reinforces its utility. During these discussions, the documentation can serve as a touchstone for evaluating what has been achieved, addressing any challenges, and updating plans as necessary. This ongoing engagement ensures that the RCA process remains a living effort rather than a one-time exercise. Progress updates can be added directly to the documentation, making it a continually evolving resource that reflects the journey toward resolution.

The benefits of RCA documentation extend beyond solving the immediate problem. It can serve as a valuable learning tool for training new team members or as a reference for tackling similar issues in the future. By archiving this documentation in a centralized location, organizations build a repository of knowledge that supports continuous improvement efforts. These records not only preserve the lessons learned but also demonstrate the organization's commitment to structured problem-solving and lasting change.

Ultimately, the way RCA documentation is used determines its effectiveness. By keeping it visible, engaging with it regularly, and leveraging it for broader organizational benefits, teams ensure that their problem-solving efforts translate into impactful results. These documents should not be seen as an endpoint but as a vital part of an ongoing journey toward excellence.

Chapter 9: Common RCA Challenges and How to Overcome Them

Introduction

Root Cause Analysis (RCA) is a transformative methodology for identifying and addressing problems at their origin. It is a cornerstone of effective problem-solving, enabling organizations to achieve long-lasting solutions rather than temporary fixes. However, while RCA holds immense potential, the path to success is rarely without obstacles. Many teams encounter barriers such as resistance to change, limited access to data, cognitive biases, or a tendency to assign blame. These challenges, if unaddressed, can undermine the effectiveness of the entire RCA process.

Why are these challenges so pervasive? RCA demands more than technical knowledge; it requires collaboration, open-mindedness, and a willingness to question deeply held assumptions. The human dynamics of problem-solving—different perspectives, competing priorities, and fear of repercussions—can create friction that derails even the most well-intentioned efforts. Left unchecked, these issues can turn RCA into an exercise in frustration, yielding superficial answers that fail to address the root causes.

Yet, for every challenge, there is an opportunity. Teams that learn to recognize and navigate these obstacles not only improve their problem-solving capabilities but also strengthen their culture of collaboration and innovation. By turning challenges into learning moments, organizations can unlock the full power of RCA.

Avoiding the Blame Game

Shifting the Focus from “Who” to “What” and “Why”

One of the most common pitfalls in Root Cause Analysis (RCA) is the tendency to assign blame. When problems arise, the instinct to find a scapegoat can dominate the discussion, leading to defensive behavior, fractured teams, and shallow solutions. Blame not only undermines trust but also shifts attention away from understanding the systemic factors that allowed the issue to occur.

To make RCA effective, leaders must actively steer the conversation away from blame and towards constructive inquiry. This requires a conscious effort to reframe the discussion and create an environment where team members feel safe to speak openly and honestly.

Reframe Questions to Focus on Systems, Not People

The language we use during RCA sessions sets the tone for the entire process. Questions that focus on "who" naturally lead to finger-pointing and defensiveness. Instead, framing questions around "what" and "why" encourages objective exploration of the problem.

- Instead of asking: “Who made this mistake?”
- Ask: “What conditions allowed this mistake to happen?” or “Why did this error occur?”

This subtle shift in language redirects attention to the processes, systems, and conditions that contributed to the issue. For example, if a shipping error occurred, asking, “What in our shipping process might have caused this?” invites a deeper examination of procedures, tools, or training gaps, rather than placing blame on an individual employee.

Highlighting Systemic Issues

Most issues are rarely the fault of one person. Research consistently shows that the majority of errors stem from systemic flaws—unclear workflows, outdated technology, insufficient training, or misaligned incentives.

During RCA sessions, emphasize that the goal is to uncover and address these systemic factors. Acknowledge that while human error may play a role, it is often a symptom of a deeper issue. For instance:

- A manufacturing defect might be due to inadequate machine maintenance schedules rather than operator negligence.
- A customer complaint about a delayed order could result from unclear communication protocols between sales and logistics teams rather than a single employee's oversight.

By focusing on systemic issues, teams not only address the immediate problem but also strengthen the overall process to prevent similar issues in the future.

Promoting Team Accountability

RCA should be framed as a collective effort to understand and improve. When everyone feels accountable for the outcome, the conversation shifts from blame to collaboration. Here is how to foster team accountability:

1. **Create Shared Ownership:** Begin the session by reminding participants that the purpose of RCA is to improve the organization's processes, not to assign blame. Stress that everyone in the room has a role in both identifying the problem and implementing the solution.

2. **Acknowledge Contributions:** Recognize that each person brings unique insights to the table. For example, a front-line worker may provide crucial details about day-to-day operations that management is unaware of, while a manager might offer a broader strategic perspective.
3. **Celebrate Transparency:** Reward honesty and openness. If someone points out a flaw in their own workflow or acknowledges a mistake, praise their willingness to contribute to the improvement process. This reinforces the idea that the session is about learning, not punishment.
4. **Encourage Neutral Language:** Use terms like "we," "our processes," or "the system" to emphasize collective responsibility. Avoid phrasing like, "You didn't follow the protocol," which singles out individuals and invites defensiveness.

Building a Blame-Free Culture

Avoiding finger-pointing in RCA is not just about improving individual sessions—it is about fostering a blame-free culture within the organization. Here are some strategies to embed this mindset into daily operations:

- **Establish Psychological Safety:** Create an environment where employees feel safe admitting mistakes or pointing out potential issues without fear of repercussions. Leaders play a critical role here by modeling humility and openness.
- **Focus on Learning:** Frame every RCA session as an opportunity to gain experience and improve. Highlight how past RCAs have led to meaningful changes, reinforcing the value of the process.
- **Use Neutral Facilitation:** Consider having an external facilitator or neutral party lead the RCA session to ensure objectivity. This can help reduce perceptions of bias or favoritism.

- **Tie Outcomes to Improvement:** Clearly demonstrate how the insights gained from RCA lead to better results for everyone involved. For instance, show how changes made after a shipping error resulted in fewer customer complaints and reduced workload for the team.

A Case in Point: Turning Blame into Collaboration

A mid-sized manufacturing company experienced frequent machine breakdowns, leading to missed deadlines and frustrated employees. Initially, management blamed the operators, accusing them of neglecting routine maintenance. During an RCA session, however, the team discovered that the root cause was not negligence—it was a lack of clarity in the maintenance schedule and insufficient training on how to perform maintenance tasks effectively.

By focusing on "what" and "why" rather than "who," the team worked together to design a clear, actionable maintenance protocol. The operators were trained on the new process, and a shared digital checklist was introduced to ensure consistency. The result? A 30% reduction in machine downtime and a more engaged, empowered workforce.

Summary

Avoiding finger-pointing is essential to the success of any RCA. By focusing on "what" and "why," highlighting systemic issues, and promoting collective accountability, organizations can turn potentially divisive discussions into opportunities for growth. When teams collaborate openly and constructively, they not only solve the problem at hand but also build a stronger, more resilient organization.

Maintaining Focus

Root Cause Analysis (RCA) sessions are inherently collaborative, bringing together diverse perspectives to identify and resolve issues. However, without a structured approach, these discussions can quickly derail into blame-shifting, heated debates, or irrelevant tangents. Maintaining objectivity is critical to fostering a productive environment where the focus remains on uncovering actionable insights.

Creating an Environment for Constructive Dialogue

The foundation of a successful RCA session lies in setting a tone of mutual respect and shared purpose. Establishing a culture of curiosity, rather than judgment, empowers participants to explore problems in-depth without fear of repercussions.

Set Ground Rules for Engagement

Before diving into the analysis, it is crucial to establish clear behavioral expectations. Ground rules help prevent unproductive dynamics and keep the conversation constructive. For example:

- Respect all contributions, even if you disagree.
- Focus on the issue, not the individuals involved.
- Commit to data-driven discussions and avoid speculation.

These rules, when communicated effectively, set the stage for an inclusive and focused dialogue. Consider posting the ground rules visibly during the session as a reminder to all participants.

Set Ground Rules for Engagement

The Role of a Neutral Facilitator
Objectivity can sometimes be elusive in RCA discussions, especially when emotions run high or when a problem has caused significant disruption. This is where a skilled facilitator proves invaluable.

A facilitator's role is not to solve the problem but to steer the conversation in a way that ensures:

- Every voice is heard, including those who may feel intimidated to speak up.
- Disagreements are mediated constructively, with a focus on the problem rather than personalities.
- The team remains aligned with the RCA process and does not stray into unrelated topics.

The facilitator should remain impartial, avoiding any bias that might skew the discussion. Their ability to ask probing questions, reframe contentious points, and redirect the group toward the facts ensures that the session stays on course.

Anchoring Discussions in Data and Evidence

When opinions dominate discussions, teams risk chasing assumptions rather than addressing root causes. Sticking to the facts minimizes this risk and builds consensus around actionable insights.

Encourage participants to rely on quantitative data, such as performance metrics, error logs, or survey results, as well as qualitative insights, like customer feedback or employee observations. For example:

- Instead of debating whether delays are frequent, reference production timelines or delivery schedules to establish a clear picture.
- Use process flow diagrams or charts to visually represent problem areas, making it easier to identify gaps or inefficiencies.

To maintain focus, ask clarifying questions such as:

- "What does the data tell us about this issue?"
- "Are we addressing a symptom or the root cause?"
- "How does this observation align with the evidence we've gathered?"

Preventing Derailment

Despite the best efforts, RCA sessions can still veer off track. A facilitator can intervene gently by acknowledging the value of the input and then steering the group back to the task at hand. For example:

- "That's an important consideration, but let's circle back to the root cause we're focusing on today."
- "This is a valuable insight—let's make a note of it for a follow-up discussion so we can stay aligned with our current objective."

Fostering a Collaborative Atmosphere

When the conversation remains focused and objective, participants feel safe to share insights, challenge assumptions, and explore solutions openly. This environment does not just solve the problem at hand—it strengthens the team's collective problem-solving abilities for the future.

The Danger of Consensus Without Critical Thought

Groupthink, a phenomenon where the desire for harmony suppresses critical thought, poses a significant threat to effective Root Cause Analysis (RCA). At first glance, quick consensus may seem efficient—it streamlines decision-making and avoids conflict. However, beneath the surface, groupthink often stifles innovation, dismisses alternative viewpoints, and leads teams to settle for shallow solutions that fail to address the true root cause.

In RCA, collaboration is essential, but uncritical agreement can derail the process. When teams prioritize agreement over analysis, they risk overlooking deeper systemic issues or creative solutions that challenge the status quo. Recognizing the signs of groupthink and proactively addressing it is key to achieving meaningful outcomes.

What is Groupthink?

Groupthink is not merely collaboration gone wrong—it is a systemic breakdown of critical thinking within a group. It often emerges subtly, as team dynamics and implicit pressures push participants toward agreement.

One of the most common indicators of groupthink is pressure to conform, where individuals align with the dominant opinion to avoid conflict or gain approval. For example, if a senior leader suggests a particular solution early in an RCA session, team members may hesitate to challenge it, fearing they will be perceived as contrarian or uncooperative. This dynamic suppresses valuable perspectives and leads to decisions based more on hierarchy than merit.

Another hallmark of groupthink is the suppression of dissent. Even when participants have reservations about a proposed solution, they may self-censor, assuming their objections will disrupt the process or be dismissed. This creates an illusion of consensus, where the group appears united, but beneath the surface, critical concerns remain unvoiced.

Consider a team analyzing production delays. They might quickly agree that "training is the problem" because it feels like a straightforward answer. However, this premature conclusion could overlook systemic issues such as flawed scheduling processes, insufficient equipment maintenance, or even communication gaps. Without dissenting voices to challenge the surface-level solution, the team risks addressing symptoms rather than root causes.

The Consequences of Groupthink in RCA

When groupthink infiltrates an RCA session, the consequences can be far-reaching. Teams may implement quick fixes that fail to resolve the actual problem, wasting resources and time. Worse, these superficial solutions often lead to the recurrence of issues, eroding trust in the RCA process itself. Missed opportunities for innovation are another casualty, as unconventional ideas are dismissed or never surface.

For example, imagine a logistics team investigating delayed shipments. They quickly conclude that drivers need better training and implement a new training program. Months later, delays persist. A closer look reveals the real problem: outdated routing software that causes inefficiencies. By failing to challenge their initial assumptions, the team wasted valuable resources on a solution that did not address the root cause.

Recognizing the Signs of Groupthink

Awareness is the first line of defense against groupthink. Teams must be vigilant for signs such as overly rapid agreement, a lack of diverse input, or discomfort with disagreement. When discussions seem to flow too smoothly, it is often a red flag that critical perspectives are being overlooked.

For instance, in a high-stakes RCA session, the absence of debate should prompt facilitators to ask: Are we genuinely aligned, or are we avoiding conflict? If key stakeholders are silent, it is essential to create space for their input, ensuring that every voice is heard and considered.

Overcoming Groupthink in RCA

To counteract groupthink, teams must cultivate a culture that values critical thought and encourages constructive disagreement. The goal is not conflict for its own sake but thoughtful debate that strengthens the problem-solving process.

A neutral facilitator can play a pivotal role in guiding discussions and ensuring balance. By posing questions like "What assumptions are we making here?" or "What alternative explanations could there be?" the facilitator can steer the team away from premature conclusions and toward deeper analysis.

Structured problem-solving techniques also help mitigate groupthink. Methods like brainwriting - where participants generate ideas individually before sharing them—reduce the influence of dominant voices. Similarly, assigning a devil's advocate to challenge the group's assumptions fosters critical evaluation and uncovers potential blind spots.

In addition to techniques, fostering an environment of psychological safety is essential. Participants must feel comfortable voicing dissent without fear of criticism or exclusion. Leaders can reinforce this culture by actively soliciting alternative viewpoints and validating contributions, even when they challenge the majority opinion.

The Upside of Avoiding Groupthink

By addressing groupthink head-on, teams unlock the full potential of RCA. Discussions become richer, solutions more robust, and buy-in stronger. When diverse perspectives are actively sought, teams are more likely to identify and implement solutions that address the root cause effectively.

For example, revisiting the logistics case: Imagine if the team had used a devil's advocate to challenge the training hypothesis. They might have explored the role of routing software sooner, saving time and resources while achieving a more impactful solution.

Ultimately, avoiding groupthink is not just about solving today's problems—it is about building a culture of critical thinking and innovation that sustains long-term organizational success.

Overcoming Resistance

Resistance from team members can arise for assorted reasons—fear of being blamed, skepticism about the value of the process, or a general reluctance to participate. Root Cause Analysis (RCA), while an essential tool for solving complex problems, often challenges individuals to critically examine processes they may feel personally connected to or responsible for. This can create discomfort, defensiveness, or even outright opposition.

Understanding and addressing this resistance is vital for the success of any RCA initiative. Resistance is not just an obstacle; it is a signal. It often reveals underlying issues within the team dynamic or organizational culture, such as lack of trust, unclear communication, or previous negative experiences with similar initiatives. Acknowledging and addressing these signals head-on can turn resistance into an opportunity to build stronger engagement and foster a culture of collaboration.

The process begins with creating an environment where participants feel safe, valued, and motivated to contribute. This means establishing psychological safety—a space where individuals feel they can express concerns, admit mistakes, and share insights without fear of blame or retribution. It also involves demonstrating the value of RCA as a tool for collective improvement, rather than an exercise in fault-finding or excessive bureaucracy.

For instance, resistance may stem from past RCA sessions that failed to deliver actionable results or where follow-through was lacking. Participants might question, “Why bother? Nothing ever changes.” Addressing this skepticism upfront by explaining how this effort will be different, and providing clear commitments to action and accountability, can help rebuild trust and encourage participation.

Additionally, fostering a sense of purpose and inclusion is crucial. Team members need to see how their input contributes to solving the problem and why their unique perspectives are indispensable. When participants feel their voices are heard and their expertise is valued, they are more likely to engage meaningfully.

Identifying Resistance

Identifying resistance during an RCA session requires a mix of observation, active listening, and understanding group dynamics. Resistance can manifest in numerous ways, both overt and subtle. Here is how to recognize it effectively:

Verbal Cues

Resistance often reveals itself in what participants say—or do not say. Listen carefully to the tone and content of the discussions.

- **Dismissive Comments:** Statements like, "We've tried this before, and it didn't work," or "This isn't the real problem," may signal skepticism about the process or the problem's framing.
- **Reluctance to Engage:** Some participants may offer minimal input or respond with vague or non-committal answers, signaling a lack of interest or belief in the process.
- **Defensiveness:** Resistance can also emerge as defensive language, especially if participants feel their roles or decisions are under scrutiny. Phrases like, "That's not my responsibility," or "I wasn't involved in that part," may indicate discomfort or fear of blame.

Non-Verbal Cues

Body language often speaks louder than words, especially in group settings.

- **Avoiding Eye Contact:** Participants who are resistant may avoid direct engagement, such as looking away or focusing on unrelated materials.
- **Closed Postures:** Crossed arms, leaning back, or disengaged body language can signal resistance or discomfort.
- **Fidgeting or Distractions:** Participants might appear distracted, checking their phones, doodling, or showing signs of impatience, which may reflect disengagement.

Behavioral Patterns

Patterns in behavior over the course of the session can also signal resistance.

- **Dominating the Conversation:** Some individuals may try to steer the discussion away from the RCA process or toward unrelated topics, effectively stalling progress.
- **Withholding Information:** Resistance can also manifest as reluctance to share critical insights or data, either due to fear of blame or distrust in the process.
- **Side Conversations:** If participants engage in whispered side discussions, it can indicate they are not fully committed to the larger group's focus.

Group Dynamics

Pay attention to how the group interacts. Resistance is often contagious and can affect the overall tone of the session.

- **Lack of Collaboration:** If participants are working in silos or avoiding direct interaction with certain team members, it may signal unresolved interpersonal tensions or distrust.
- **Groupthink or Over-Agreement:** While consensus might seem positive, it can indicate that participants are avoiding conflict or critical thinking, possibly to "get it over with."

Challenges to the Facilitator

Resistance might also be directed at the facilitator or the RCA process itself.

- **Questioning the Process:** Participants might openly challenge the need for RCA or the facilitator's authority, saying things like, "Why are we doing this?" or "This won't solve anything."
- **Resistance to Tools or Techniques:** Hesitation to engage with RCA methods, such as the 5 Whys or Fishbone Diagrams, may indicate discomfort with the process or skepticism about its value.

Low Energy or Engagement

A lack of enthusiasm or energy in the room can be a subtle but telling sign of resistance.

- **Silence:** Prolonged silence or a lack of voluntary participation can indicate disinterest, fear, or lack of understanding.
- **Minimal Contributions:** Participants may only contribute when directly asked, and even then, provide short or superficial responses.

Addressing Resistance

Addressing resistance in an RCA session requires careful navigation and a thoughtful approach. Once resistance is identified, the facilitator's role is to create an environment where concerns are acknowledged, and participants feel motivated to engage in a constructive way. This involves not only recognizing the signs of resistance but also taking deliberate steps to address it effectively.

One of the most powerful tools for overcoming resistance is open-ended questioning. Questions like, "What concerns do you have about this process?" or "How do you feel about the direction we're taking?" invite participants to share their thoughts without fear of judgment. Open-ended questions are non-confrontational and allow team members to express their perspectives in their own words. By encouraging dialogue, the facilitator can uncover underlying issues that may be causing hesitation or disengagement. These questions also help redirect the conversation back to the core purpose of the session, ensuring the group remains focused on the root cause analysis.

Equally important is the ability to acknowledge and validate emotions. Participants might feel frustrated, fearful, or skeptical for assorted reasons—whether it is a past experience with an unproductive RCA session or concerns about being blamed for the problem. As a facilitator, showing empathy by recognizing these emotions can significantly lower resistance. Simple statements like, "I understand why you might feel this way," or "Your concerns are valid, and we're here to work together on this," can diffuse tension and demonstrate that everyone's voice matters. Acknowledging emotions does not mean agreeing with them; rather, it is about creating an atmosphere where individuals feel heard and respected.

Clarifying the purpose of the RCA session is another key strategy. Resistance often arises when participants do not fully understand the session's objectives or feel unsure about their role within it. Taking the time to reiterate the session's goals—focusing on improvement rather than blame—can help shift perceptions. For example, reminding the group that the RCA is a collective effort aimed at identifying systemic issues and preventing recurrence, rather than pointing fingers, can refocus their attention on the bigger picture. This reframing allows team members to see the session as an opportunity for growth rather than a critique of past actions.

Creating transparency is essential for overcoming resistance. When participants feel safe and know that their input will be valued rather than judged, they are more likely to contribute openly. Facilitators can foster transparency by setting clear expectations for behavior at the start of the session. Establishing ground rules, such as "respect all opinions" and "focus on processes, not people," reinforces the idea that the RCA is a safe and collaborative environment. Transparency also means being upfront about what the session aims to achieve and how the outcomes will be used, helping to build trust and encourage honest participation.

By combining these approaches—encouraging dialogue, validating emotions, clarifying objectives, and promoting transparency—facilitators can transform resistance into active engagement. This shift not only enhances the quality of the RCA process but also strengthens team cohesion and trust, laying the foundation for meaningful and lasting solutions.

Summary

Root Cause Analysis (RCA) is a transformative tool for addressing systemic problems, but it is not without its hurdles. This chapter explored the most familiar challenges teams face during the RCA process and provided actionable strategies to overcome them, ensuring more effective and sustainable outcomes.

One of the most pervasive issues in RCA is the tendency to assign blame, which can derail the process and stifle honest participation. Shifting the focus from "who" to "what" and "why" creates an environment where systemic issues, not individuals, are scrutinized. Facilitators play a vital role in steering discussions toward constructive problem-solving and away from finger-pointing.

Groupthink presents another significant challenge, where the desire for consensus can suppress dissent and critical thinking. By fostering an inclusive environment that values diverse perspectives, encouraging respectful debate, and designating a neutral facilitator, teams can unlock innovative solutions and avoid premature conclusions.

Resistance—whether from team members during sessions or skepticism about the RCA process as a whole—requires a tailored approach. Identifying resistance early, addressing its root causes, and creating a transparent, supportive atmosphere are critical to turning hesitant participants into engaged collaborators. This chapter highlighted the importance of validating emotions, clarifying objectives, and building trust to overcome these obstacles.

Finally, the chapter emphasized the importance of focusing conversations on objective data and evidence. Structured discussions grounded in facts help maintain the integrity of the RCA process and prevent it from being derailed by individual opinions or tangential debates.

By understanding and addressing these challenges, leaders can transform RCA sessions from potential points of conflict into opportunities for collective growth and problem-solving. This proactive approach not only strengthens the process itself but also fosters a culture of accountability, collaboration, and continuous improvement within the organization.

Chapter 10: Quantifying and Shortlisting Potential Solutions

Introduction:

After facilitating an RCA session, it is common to uncover a wide array of potential improvements. In one particularly complex and protracted session I led, we identified over fifty opportunities for change. While this abundance of options may seem like a win, it also presents a challenge: how do you decide which solution to pursue? Each potential improvement carries its own set of costs, risks, and benefits, making the process of narrowing them down anything but straightforward.

Resources are often limited, and the pressure to make the right decision can feel overwhelming. Relying on instinct or incomplete information can lead to wasted time and effort, or worse, solving the wrong problem entirely. A thoughtful, structured approach can help turn what feels like a daunting task into a clear path forward. By evaluating options systematically and basing decisions on evidence, leaders can ensure that the solutions they choose will truly address the issues at hand and deliver lasting results.

This chapter explores the process of quantifying potential solutions, from assessing feasibility and conducting cost-benefit analyses to evaluating risks and defining success metrics. By following these steps, you will learn how to systematically evaluate your options, prioritize effectively, and create a shortlist of solutions that deliver the greatest impact with the resources available.

Assessing Feasibility: Can We Implement This Solution?

Assessing feasibility is the first step in turning a potential solution into actionable change. Not every promising idea is immediately practical, and some may require adjustments before they can be implemented. By carefully evaluating feasibility, you can filter out solutions that are currently unworkable while identifying opportunities to refine or scale ideas to make them achievable.

Evaluating feasibility is not just about ruling out options—it is about finding the sweet spot between effort and reward. High-effort solutions with substantial rewards may justify the investment, particularly for strategic priorities. Conversely, low-impact solutions that demand significant resources can often be eliminated early.

The key is to approach feasibility as a dynamic process. By leveraging stakeholder input, exploring phased implementation, and remaining flexible in resource allocation, you can turn even seemingly challenging solutions into actionable plans.

What Makes a Solution Feasible?

Feasibility hinges on several key factors, including resource availability, time constraints, organizational capabilities, financial considerations, and the balance of effort versus reward. A feasible solution is one that can be implemented with reasonable effort relative to the impact it delivers.

- Effort vs. Reward: Does the anticipated outcome justify the effort required? A high-effort solution may be worth pursuing if it delivers transformative results, but low-impact solutions rarely merit significant investment.
- Resource Availability: Do you have the necessary financial, human, and material resources? A new process or system might promise significant improvements, but without the required budget or skilled personnel, it could stall before implementation begins.
- Time Constraints: Can this solution be implemented within the necessary timeframe? Urgent problems may require solutions with shorter lead times, even if more comprehensive options offer greater long-term benefits.
- Organizational Capabilities: Does your team have the expertise, technology, and infrastructure to execute the solution? If not, consider whether upskilling, hiring, or outsourcing could bridge the gap.
- Financial Considerations: Is the cost of the solution justifiable given your organization's financial priorities? Even feasible solutions may need to be deferred if their costs outweigh current budget constraints.

Scenarios Where Infeasible Solutions Can Become Feasible

A solution initially deemed infeasible may become viable with adjustments or creative problem-solving. Consider the following scenarios:

- Resource Reallocation: Suppose a solution requires more funding than initially budgeted. By reallocating resources from lower-priority projects, you may free up the necessary budget to pursue the solution.
  Example: A manufacturing facility identifies the need for new equipment to address bottlenecks but lacks the upfront capital. By temporarily pausing a non-essential expansion project, funds can be redirected to the immediate need.

- External Funding or Partnerships: External funding sources, such as grants or strategic partnerships, can transform an otherwise out-of-reach solution into a feasible one.
  Example: A municipality considering a large-scale recycling program partners with a private waste management company to share costs and resources, making the initiative achievable.

- Incremental Implementation: Breaking a complex solution into smaller, manageable steps can make it more feasible.
  Example: Instead of implementing a full system overhaul, an IT department might start with a pilot program for a single team, allowing the solution to prove its value before scaling up.

The Role of Stakeholder Input

Engaging stakeholders during the feasibility assessment is critical to grounding the process in practical realities and ensuring organizational alignment. Stakeholders, particularly those directly involved in day-to-day operations, bring valuable insights into potential obstacles or bottlenecks that may not be immediately apparent to leadership. Their input helps create more realistic and actionable plans by highlighting operational nuances and constraints. Moreover, involving stakeholders early fosters a sense of ownership and builds buy-in, increasing the likelihood of successful implementation. When stakeholders feel their perspectives are heard and valued, they are more likely to actively support the chosen solution and contribute to its success. This collaborative approach ensures that feasibility assessments are comprehensive and the resulting action plans are both practical and achievable.

The Balancing Act

Feasibility assessments require balancing ambition with practicality. While it is important to stretch for impactful solutions, overcommitting can lead to frustration and failure. By thoroughly evaluating resources, timelines, financial considerations, risks, and stakeholder input, you will ensure that the solutions you pursue are both achievable and impactful. And when a solution feels just out of reach, creative adjustments can often bring it back into the realm of possibility.

Cost-Benefit Analysis: Measuring Financial Impact

Once feasibility is determined, the next step in narrowing down potential solutions is to analyze their financial implications. Cost-Benefit Analysis (CBA) is a cornerstone of effective decision-making, providing a systematic way to evaluate the trade-offs between the costs of implementing a solution and the benefits it is expected to deliver, the investment (ROI). This process not only ensures the efficient use of resources but also helps align decisions with strategic priorities.

Why Cost-Benefit Analysis Matters

At its core, a CBA ensures that resources are directed toward solutions that maximize value. Without this step, organizations risk pursuing options that drain resources without providing meaningful improvements. CBA provides clarity by quantifying both tangible and intangible aspects, offering decision-makers a holistic view of each solution's financial impact.

Consider a manufacturing company deciding between two potential solutions for reducing production downtime: upgrading existing machinery or replacing it with state-of-the-art automated equipment. A robust CBA would weigh the upfront and operational costs of both options against the expected gains in productivity and efficiency. This ensures that the chosen solution delivers the best value for the investment.

Components of a Cost-Benefit Analysis:

Breaking Down the Costs and Benefits

Costs:
Costs can be categorized into direct, indirect, and opportunity costs.

- Direct Costs include tangible expenditures such as equipment, labor, or software. For example, implementing a new software system might require purchasing licenses, hiring consultants, and training staff.
- Indirect Costs involve less obvious expenditures, like temporary operational disruptions or the learning curve associated with a new process.
- Opportunity Costs represent the benefits forgone by choosing one solution over another. For instance, allocating resources to a major system overhaul might delay other important initiatives.

Benefits:
Benefits encompass measurable gains such as cost savings, revenue increases, and productivity improvements, along with less tangible rewards like enhanced reputation or employee satisfaction.

- Tangible Benefits: These are quantifiable outcomes like reduced operational costs, increased production output, or enhanced customer retention rates.
- Intangible Benefits: While harder to measure, these can include improved team morale, stronger customer relationships, or enhanced brand reputation. For instance, a solution that improves service delivery times might boost customer loyalty, a benefit that compounds over time.

Example:

A retail company is deciding between upgrading its existing point-of-sale (POS) system or implementing a new, cloud-based solution. Through a CBA, they discover:

- Upgrading the current system would cost $50,000 and improve transaction speed by 10%.
- Implementing a new cloud-based system would cost $100,000, but improve transaction speed by 25%, allow for seamless updates, and improve customer data analysis capabilities—leading to a projected 15% increase in customer retention.

While the new system is more expensive upfront, the projected long-term gains in customer retention and data insights provide a better ROI.

Risk vs. Reward: Weighing Potential Risks Against Benefits

All solutions carry a degree of risk, and it is important to assess these risks in relation to the potential rewards. A high-reward solution may seem attractive, but the associated risks could threaten your business if they are not effectively managed. Effective risk analysis ensures you are not blindsided by unintended consequences.

Types of Risks to Consider:

1. **Operational Risks:** Will the solution disrupt your day-to-day operations during implementation? For example, replacing a major IT system could cause downtime or data loss if not managed correctly.
2. **Financial Risks:** Could the solution overrun its budget, or could the financial returns be less than expected? Unforeseen costs like increased maintenance or underperformance can result in financial strain.
3. **Market Risks:** Will this solution change your market position or expose you to new competitors? A solution that seems internally sound may shift customer perceptions or market dynamics in unforeseen ways.

4. **Reputational Risks:** How will the solution affect your company's reputation if it fails? Customer-facing solutions need to be particularly careful here—any misstep could lead to customer dissatisfaction, lost business, or bad press.

Risk Mitigation Strategies:

1. **Pilot Testing:** Before full implementation, consider rolling out the solution on a small scale to evaluate its feasibility and impact. This will give you a clearer picture of any unexpected issues and allow you to refine the process.
2. **Contingency Planning:** Develop backup plans for high-risk solutions. If your solution involves major system changes, ensure you have a recovery plan in case things do not go as expected.

Example:

A healthcare company is considering implementing artificial intelligence (AI) to assist in diagnostics. While the potential reward is improved diagnostic accuracy and faster patient care, the risks include compliance with regulatory requirements, patient privacy concerns, and the cost of training doctors to use the new system.

The company decides to pilot the AI tool in a small number of clinics to assess its impact and gather feedback before committing to a broader rollout. This risk mitigation strategy ensures they can adjust before facing larger-scale challenges.

Metrics for Success: How Will We Measure the Impact of Each Solution?

Once the feasibility and cost-benefit aspects of each solution have been assessed, it is essential to define the metrics for success. These metrics allow you to measure whether the solution delivers the expected results once implemented. Clear metrics ensure accountability and provide an objective way to evaluate performance post-implementation.

Key Performance Indicators (KPIs):

1. **Time Savings:** This is often a key metric, especially for operational problems. How much time does the solution save, and how does that translate into efficiency gains?
2. **Cost Savings:** Direct savings can include reductions in labor, materials, or maintenance costs. Cost savings may also come from reducing downtime or preventing future issues.
3. **Quality Improvements:** If solutions are aimed at improving product or service quality, track relevant metrics like defect rates, customer complaints, or product returns.
4. **Customer and Employee Satisfaction:** How does the solution impact the people who engage with it? Surveys, feedback mechanisms, and retention rates provide insight into both customer and employee satisfaction levels.
5. **Revenue Impact:** Whether the solution leads to increased sales, improved customer.

Example:

A logistics company implements route optimization software to reduce fuel costs and improve delivery times. To measure the success of the implementation, they set KPIs around:

- **Fuel cost reduction:** aiming for a 15% decrease in fuel expenses.
- **Delivery speed:** targeting a 10% improvement in average delivery times.
- **Customer satisfaction:** aiming to increase positive delivery reviews by 20%.

These KPIs give the company tangible goals to track and evaluate whether the solution is delivering on its promises.

How to Conduct a Cost-Benefit Analysis

Performing a Cost-Benefit Analysis is as much an art as it is a science. It begins with carefully identifying and quantifying all costs and benefits associated with a potential solution. This involves diving into both direct expenses, like materials or labor, and indirect costs, such as temporary disruptions or the learning curve for employees. While tangible costs are often easier to calculate, it is equally important to consider the less obvious aspects—like the opportunity costs of choosing one solution over another.

Quantifying costs and benefits require a systematic approach to ensure accuracy and reliability. For costs, this might involve gathering historical data, consulting with vendors, or engaging subject matter experts to estimate expenses. For benefits, the process often includes projecting gains based on key performance indicators (KPIs) such as revenue increases, efficiency improvements, or customer satisfaction scores. It is also important to differentiate between tangible benefits, like reduced operational costs, and intangible ones, such as improved team morale or brand reputation. Assigning realistic monetary values to both types of benefits can help create a more balanced and comprehensive analysis.

Once you have compiled a comprehensive list, the next step is to evaluate how costs and benefits unfold over time. For instance, some solutions might require significant upfront investments but deliver benefits steadily over years, while others might yield quicker wins but offer less long-term value. Using tools like Net Present Value (NPV) can help you compare these timelines effectively, ensuring that long-term gains are not overshadowed by immediate expenses.

It is also critical to incorporate the variability and risks inherent in any solution. Not every outcome is guaranteed, so performing a sensitivity analysis can be a game-changer. This approach allows you to assess different scenarios—what happens if costs rise unexpectedly, or if benefits are delayed? By exploring these "what-if" scenarios, you can better understand how robust a solution is under varying conditions and avoid unpleasant surprises down the road.

A successful Cost-Benefit Analysis is not just about numbers on a page—it is about telling the story behind the data. Why does one solution make sense over another, and how do these choices align with the organization's broader goals? By connecting the financial analysis with strategic priorities, leaders you will be better equipped to select solutions that deliver meaningful and lasting results.

Prioritizing Solutions Based on Impact (with 2x2 Effort vs. Impact Matrix)

Once you have evaluated your potential solutions based on feasibility, cost, risk, and KPIs, it is important to prioritize them. The Effort vs. Impact Matrix, commonly represented as a 2x2 grid, is an effective way to do this. It helps you identify which solutions offer the most impact for the least amount of effort, allowing you to focus on initiatives that will deliver the greatest value with the least resource expenditure.

Understanding the 2x2 Effort vs. Impact Matrix:

The matrix divides potential solutions into four quadrants based on two criteria:

- **Effort**: How much time, money, and resources will be required to implement the solution.
- **Impact**: The anticipated positive outcome of the solution, such as cost savings, improved efficiency, or customer satisfaction.

The matrix looks like this:

Anticipated Benefit	High impact	Major Projects	Quick Wins
	Low Impact	Avoid	Low Priority
	2 x 2 Matrix	High Effort	Low Effort
		Effort required to Implement	

Categories of Solutions in the Matrix:

1. **Quick Wins (Low Effort, High Impact):** These are the first solutions to pursue. They provide significant value without requiring substantial resources, offering the highest return on investment in the shortest amount of time. These are typically smaller fixes that can be implemented quickly and immediately improve operations. Example: A minor software upgrade that reduces the manual data entry time for employees, resulting in a quick productivity boost.

2. **Major Projects (High Effort, High Impact):** While these solutions are resource-intensive, their potential to generate long-term benefits makes them worthwhile. These are strategic initiatives that may take longer to implement but have a significant positive impact on business performance. Example: A complete overhaul of your enterprise resource planning (ERP) system. This project will require months of implementation, training, and significant financial investment, but it has the potential to streamline operations and deliver significant gains in productivity.

3. **Low-Priority Tasks (Low Effort, Low Impact):** These are the tasks that, while easy to implement, do not offer much in terms of results. These can be pursued if resources are available, but they are not critical. Example: Minor workflow tweaks that save small amounts of time but do not drastically change overall performance.
4. **Avoid (High Effort, Low Impact):** These are solutions that require substantial resources but provide limited benefits. These should be avoided unless absolutely necessary. Example: Implementing a complex and expensive tool that only marginally improves one area of your operations.

How to Use the 2x2 Matrix:

1. **Plot Your Solutions:** After assessing feasibility, costs, risks, and metrics, categorize each solution into the appropriate quadrant based on its expected effort and impact.
2. **Prioritize Quick Wins:** Focus first on solutions that fall into the Quick Wins quadrant. These offer the most value for the least effort and will deliver fast, visible improvements.

3. **Strategize for Major Projects:** High-effort, high-impact projects should be strategically planned and resourced. While they take more time to implement, they are likely to yield long-term success.
4. **Reassess Lower-Impact Solutions:** For solutions in the Low-Priority or Avoid quadrants, reassess whether they are worth the investment. If resources are scarce, it is better to concentrate on solutions that provide meaningful returns.

Example:

A marketing team is trying to decide how to allocate resources for the next quarter. They have identified the following potential solutions to improve customer acquisition:

- **Low Effort, High Impact**: Creating automated email campaigns targeting warm leads, expected to boost conversions by 15%.
- **High Effort, High Impact**: Redesigning the website to improve user experience, which requires a significant investment but could increase customer retention by 25%.
- **Low Effort, Low Impact**: Updating the FAQ section on the website, which is easy to do but will not lead to significant changes in customer acquisition.
- **High Effort, Low Impact**: Launching a new social media platform with no proven success in their target market, which would require high marketing and content creation costs but is unlikely to deliver results.

Using the 2x2 matrix, the team prioritizes the email campaigns and website redesign, deferring the other options for later consideration.

Working Example

A production facility was having consistent poor performance on one of their production lines. After doing the root cause analysis, they came up with four potential solutions to reduce the impact. Those were put into a spreadsheet and evaluate it on opportunity type, savings, cost implement, and net benefit.

ID	Opportunity Title	Opportunity Type (Efficiency, Cost Reduction, EHS, Morale)	Validated Savings	Cost To Implement	Net Benefit	Non-Cost Benefit Value Scale: 1 = Little Benefit, 5 = Great Benefit	Impact on Performance/$ Scale: 1 = Low, 5 = High	Ease of Implementation Scale: 1 = Difficult, 5 = Easy
1	Change rollers	Performance	$ 10,000	$ 5,000	$ 5,000	4.0	4.0	4.0
2	Rebuild boiler	Reliability	$ 15,000	$ 150,000	$ (135,000)	1.5	1.5	1.0
3	Change drive belts	Efficiency	$ 7,500	$ 7,500	$ -	2.5	5.0	2.5
4	Replace sensors	Efficiency	$ 25,000	$ 15,000	$ 10,000	4.5	4.5	3.0

This information is plotted on a 2 x 2 matrix:

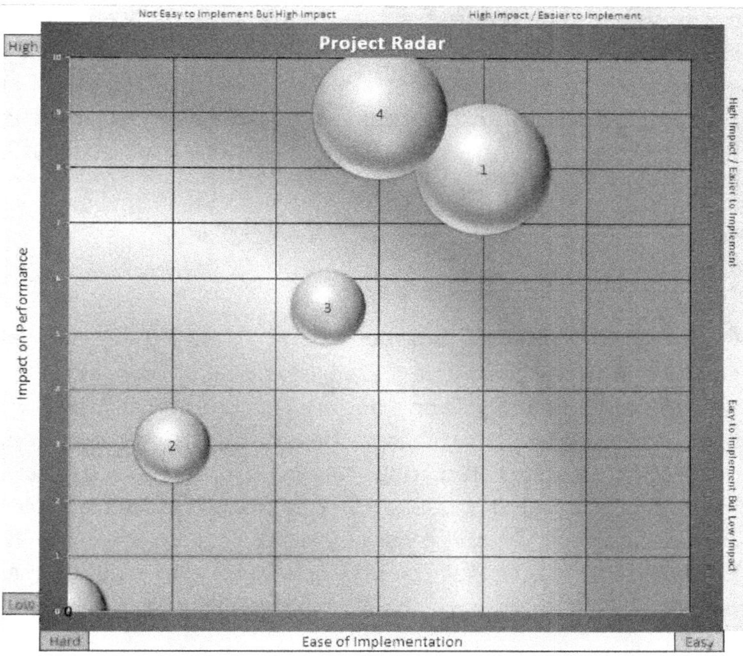

Projects 1 and 4 were in the High Impact/Easier to Implement quadrant where project 2, rebuilding the boiler fell into the Low Impact/Hard to Implement quadrant. Looking at the financial benefits, Although project 4, replacing the sensors cost $25,000 to implement, the calculated return was greater that changing the rollers and was implemented. After implementation, the actual benefit derived exceeded expectations.

Final Shortlisting: Making the Decision

After completing the evaluation process, the final step is to develop a shortlist of the most viable solutions. The shortlisted options should be those that provide the highest overall value, balancing feasibility, cost-effectiveness, risk, and impact. It is essential to bring all relevant stakeholders into this decision-making process to ensure that everyone is aligned, and that the solution has the necessary support for successful implementation.

Summary

Quantifying potential solutions and creating a focused shortlist is more than just a step in the RCA process; it is a bridge between analysis and action. This chapter underscores the importance of moving beyond gut feelings and embracing a structured, data-driven approach to decision-making. By carefully assessing feasibility, conducting cost-benefit analyses, weighing risks, and utilizing tools like the 2x2 Effort vs. Impact Matrix, organizations can focus their resources on solutions that deliver the greatest value.

The process of quantifying solutions ensures that leaders make choices rooted in clarity and confidence, setting the stage for impactful results. However, it is not just about numbers—it is about aligning decisions with strategic goals, fostering cross-functional collaboration, and maintaining accountability throughout the process. Whether pursuing a quick win or planning a transformative project, the insights gained during this phase empower teams to act decisively and effectively.

The value of quantified solutions lies in their ability to drive lasting change. By choosing solutions that are practical, impactful, and sustainable, leaders can solve not just the immediate problem but also build a foundation for continuous improvement. This structured approach to decision-making is essential for any organization striving to remain competitive and achieve operational excellence in an ever-evolving business landscape.

Chapter 11: Turning Insight into Action – Crafting an Effective Action Plan

Introduction

Defining the root cause of a problem is a significant milestone in the problem-solving journey, but the challenge lies in translating that insight into meaningful, measurable action. An effective action plan serves as the bridge between understanding a problem and implementing solutions that deliver lasting results. Without a structured roadmap, even the most thorough Root Cause Analysis (RCA) risks falling short, leaving the organization with unresolved issues, and missed opportunities.

Action planning is more than a checklist; it is a strategic framework that aligns resources, timelines, and accountability to ensure success. It is about taking the momentum from your RCA session and channeling it into a focused effort that tackles the problem head-on while fostering team collaboration and engagement. This step is critical, not just to resolve the immediate issue but to embed lasting improvements into the organization's processes.

In this chapter, we will explore how to craft an effective action plan that turns RCA findings into actionable steps. From defining clear objectives and prioritizing tasks to assigning ownership and celebrating milestones, this chapter will provide a practical guide to moving from insight to impact. Along the way, you will learn how to balance ambition with realism, inspire your team to act, and create a culture where change is not only achievable but sustainable.

Creating the Action List

An effective action list serves as the foundation for translating plans into tangible outcomes. It provides the structure teams need to move forward, ensuring no details are overlooked and responsibilities are clearly defined. Developing an action list is not just about writing tasks—it is about prioritizing efforts, fostering accountability, and making sure every step aligns with the overarching goal.

Defining the Action Items

At its core, an action list is a breakdown of the steps required to achieve the desired outcome. Each item should be specific, actionable, and directly tied to resolving the root cause of the problem. Vague tasks like "fix the process" do not provide the clarity needed to drive results. Instead, focus on defining actions that clearly state what needs to be done, such as "streamline data entry by implementing automated software."

For example, if a team is tasked with reducing equipment downtime, the action list might include:

- Conducting a root cause analysis to identify failure points.
- Scheduling regular maintenance intervals.
- Training operators to identify early warning signs of equipment wear.

Each action should represent a logical step toward the goal, ensuring that progress is measurable and methodical.

Assigning Responsibility

A clear action list is only effective if responsibility for each task is explicitly assigned. Ambiguity in ownership can lead to delays, duplication of effort, or tasks falling through the cracks. For every action item, designate a single person or team accountable for its completion. This individual does not necessarily have to perform the task alone but must oversee its execution and ensure it aligns with the timeline and objectives.

For instance, a project involving workflow automation might assign the IT team to implement the software, while the operations team ensures seamless integration into daily activities. By having a named point of accountability for each task, progress becomes easier to track, and potential bottlenecks can be identified and addressed swiftly.

Providing Resources and Training

Assigning tasks is only part of the equation; teams must also have the skills and tools required to execute them effectively. Leaders should assess whether individuals need additional training or resources to succeed. For example, if a task involves implementing new software, ensure that the team is trained to use it efficiently before deployment. Investing in upskilling employees not only ensures the action list is completed successfully but also enhances the team's capabilities for future initiatives.

Sequencing and Dependencies

Not all actions can or should be tackled simultaneously. Sequencing tasks logically and identifying dependencies is crucial for ensuring smooth execution. For instance, installing new equipment may require that training be completed first, or creating a new workflow might hinge on software updates. By mapping out the sequence of tasks and highlighting dependencies, teams can avoid delays and maintain momentum.

A practical approach is to use tools like Gantt charts or project management software to visualize the timeline and dependencies, ensuring that resources are allocated efficiently, and potential conflicts are resolved proactively.

Prioritizing for Impact

While all tasks on the action list are important, some will have a greater impact on the overall goal than others. Begin by prioritizing tasks that address the most pressing issues or offer quick wins. For example, if the goal is to reduce defects in production, start with tasks that immediately improve quality control processes before tackling longer-term structural changes. Prioritization ensures that progress is made steadily, keeping the team motivated and the project on track.

Communicating the Action Plan

Once the action list is created, share it with all stakeholders. Clear communication ensures alignment, fosters collaboration, and allows for input that could refine the plan further. A well-communicated action plan acts as a unifying document, helping everyone understand their role and how their efforts contribute to the larger goal.

Keeping the List Dynamic

Finally, remember that an action list is not set in stone. As circumstances evolve or new insights emerge, adjust the tasks to reflect changing priorities or conditions. Regularly reviewing and updating the action list ensures it remains relevant and effective throughout the implementation process.

By creating a detailed and well-thought-out action list, teams have a roadmap for turning insights into action. A strong list empowers individuals, fosters accountability, and ensures progress is methodical, measurable, and sustainable. Whether addressing a small issue or tackling a major challenge, the action list is where analysis transforms into impactful, real-world solutions.

Crafting an Effective Action List: The Four Essentials

An action list is the backbone of any successful plan. It translates high-level goals into specific, measurable tasks and provides a roadmap for teams to follow. For an action list to work effectively, it must include four essential components: what the action items are, who owns them, when they are due, and their current status. Incorporating these elements ensures clarity, accountability, and progress tracking.

What the Action Items Are

The action list must clearly articulate the specific tasks required to achieve the goal. Each action item should be descriptive enough to avoid ambiguity while remaining concise.

Example:

Instead of writing "Fix production issue," write "Repair faulty conveyor system on Line B to reduce downtime."

A detailed action item ensures everyone understands exactly what needs to be done, minimizing delays caused by confusion or miscommunication.

Who Owns the Actions

Every action item must have a clearly designated owner—someone responsible for seeing the task through to completion. Ownership fosters accountability and ensures no task falls through the cracks.

Example:

Assign the maintenance manager as the owner of "Repair faulty conveyor system on Line B."

When assigning ownership, ensure the individual has the authority, resources, and skills to complete the task. If they do not, provide the necessary support, such as training or additional resources.

When the Tasks Are Due

Setting specific due dates creates urgency and helps maintain momentum. Deadlines should be realistic yet firm, balancing the need for timely action with the complexity of the task.

Example:

"Repair faulty conveyor system on Line B" could have a due date of "December 15th."

Establishing clear deadlines also allows for better planning and helps prioritize tasks when multiple actions compete for attention.

What the Current Status Is

Tracking the status of each action item ensures visibility into progress and highlights any potential bottlenecks. Common status categories include:

- Assigned
- Not Started
- In Progress
- Completed
- Overdue
- Blocked

Including a status column in the action list allows leaders to monitor progress immediately and address issues before they escalate.

Enhancing the Action List: Nice-to-Have Elements

While the four essentials provide a solid foundation, additional elements can further enhance the action list's effectiveness:

Detailed Description

Adding context or specific details about each action item can help team members understand the task's scope and importance. This is especially useful for complex or multi-step actions.

Example:

"Repair faulty conveyor system on Line B to address downtime and prevent further production delays."

Assigned Date

Tracking when the task was assigned provides valuable context for progress reviews. It helps identify whether delays are due to late starts or prolonged execution.

Example:

Assigned Date: "November 25th."

Notes Section

A notes section allows for dynamic updates, such as recording obstacles encountered, additional resources required, or interim progress.

Example Notes:

"Spare parts on order, expected delivery December 10th."

Example Action List in Practice

Here is how an action list might look in action for a manufacturing team addressing equipment downtime:

No.	Action Items	Owner	Date Created	Due Date	Status	Notes
1	Install visual management board	[illegible]	12/15/2024	1/6/2025	In Progress	Board ordered. Exp. delivery Jan [illegible], 2025
2	Create production metrics for line 3	[illegible]	12/15/2024	1/6/2025	Complete	
3	Order belt 3 sensors	Ling	12/15/2024	1/6/2025	Complete	Expected Dec 27, 2024
4	Install belt 3 sensors	Ling	12/15/2024	1/10/2025	Pending	Scheduled install Jan 6, 2025
5	Visual management training	[illegible]	12/15/2024	1/6/2025	In Progress	Scheduled week of Jan 5th
6	Project status review meeting	[illegible]	12/15/2024	1/15/2025	Not started	

Why This Matters

An action list with these components is more than a to-do list —it is a structured roadmap for execution. By clearly defining what, who, when, and status of each task and supplementing it with contextual details, leaders can maintain focus, foster accountability, and ensure steady progress toward achieving their goals.

Leveraging the SMART Framework to Strengthen Your Action Plan

The SMART framework—Specific, Measurable, Achievable, Relevant, and Time-bound—is a powerful tool to create focused, actionable goals that drive progress. When applied to action planning, SMART goals clarify what needs to be done, ensure alignment with larger objectives, and provide a way to measure success.

Specific: Define the What, Why, and How

A goal should be clear and precise, leaving no room for ambiguity. Specificity ensures everyone understands exactly what is being addressed, why it matters, and how it will be achieved.

- **Example:** Instead of saying, "Improve production," a SMART goal would specify, "Increase production of Product A by 10% to meet customer demand."
- **Why It Matters:** Specific goals provide a concrete focus, making it easier to create actionable steps and avoid misinterpretation.

Measurable: Quantify Success

Measurable goals include criteria to track progress and assess success. Metrics should be objective and relevant to the desired outcome.

- **Example:** "Reduce machine downtime by 20% over the next three months by improving maintenance schedules."
- **How to Measure:** In this example, downtime logs and maintenance reports can be used to track progress and ensure the target is achieved.

- **Why It Matters:** Measurability allows teams to monitor performance, adjust strategies as needed, and celebrate milestones along the way.

Achievable: Balance Ambition with Realism

Goals should stretch the team's capabilities but remain within the realm of possibility given current resources, expertise, and constraints.

- **Example:** "Train 100% of the operations team on the new quality control system by the end of Q1."
- **How to Assess Feasibility:** Consider factors like training schedules, resource availability, and team capacity to ensure the goal can realistically be accomplished.
- **Why It Matters:** Unrealistic goals can lead to frustration and disengagement, while achievable goals foster motivation and confidence.

Relevant: Align with Broader Objectives

A goal must align with organizational priorities and contribute to overarching success. Relevance ensures the effort invested in achieving the goal directly supports larger strategic outcomes.

- **Example:** "Develop a supplier evaluation process to improve material quality and reduce defects by 15%."
- **Why It Matters:** If improving material quality is a critical priority for reducing production errors, this goal aligns directly with broader operational objectives.

Time-bound: Set Deadlines for Accountability

Every SMART goal needs a specific timeline. Deadlines create urgency, encourage accountability, and help prioritize tasks effectively.

- **Example:** "Complete the implementation of the new inventory management system by March 15th."
- **Why It Matters:** Time constraints help prevent procrastination and maintain momentum. They also provide a clear timeframe for evaluating success.

Working Example

Imagine a scenario where a manufacturing team wants to improve production efficiency. Using the SMART framework, their goal might be:

- **Specific:** "Optimize the production line for Product B to reduce cycle time."
- **Measurable:** "Achieve a 25% reduction in cycle time."
- **Achievable:** "Reallocate existing resources and train the workforce to implement the new workflow."
- **Relevant:** "Supports the company's strategic goal of increasing throughput by 20% this year."
- **Time-bound:** "Complete the optimization process by June 30th."

By applying the SMART framework, the team transforms a vague aspiration into a clear, actionable goal with defined parameters for success.

The Power of Incremental Change

When it comes to addressing complex problems, the temptation to seek a sweeping, transformative solution is natural. However, the power of incremental change often lies in its ability to deliver faster, more manageable wins while reducing risks associated with large-scale changes. Incremental improvements are not only easier to implement but also provide the opportunity to assess, adapt, and refine approaches before scaling them further.

Why Incremental Change Matters

1. **Building Momentum**: Tackling smaller, achievable changes creates a sense of progress and accomplishment. This momentum motivates teams to continue making improvements and fosters a positive mindset toward change.
2. **Lowering Resistance to Change**: Large-scale changes often provoke anxiety and resistance among employees. Incremental improvements, on the other hand, are less disruptive and easier for teams to embrace.
3. **Reducing Risk**: Small changes allow teams to identify potential pitfalls early, minimizing the impact of errors or setbacks. If something does not work, it is easier to pivot without significant loss of time or resources.
4. **Continuous Improvement Culture**: Incremental change reinforces a mindset of ongoing progress, helping to embed continuous improvement into the organization's DNA.

Applying Incremental Change

Incremental change works best when it aligns with the broader goals of the action plan while focusing on practical, immediate wins. Here is how to apply it effectively:

- **Start Small**: Identify a specific area where a small improvement can make a noticeable difference. For example, streamline a single step in a production process or reduce redundant approvals in a workflow.
- **Test and Learn**: Treat each incremental improvement as a mini experiment. Gather feedback and measure results to determine what works and what does not. This iterative approach allows for adjustments and refinements before scaling up.
- **Stack Improvements**: Build on each success by identifying the next incremental change. Over time, these incremental changes compound, leading to significant overall progress.

Example: Automating Manual Processes

Imagine a finance department where employees spend hours manually reconciling accounts each week. Instead of attempting to overhaul the entire reconciliation process, start by automating just one or two repetitive tasks, such as matching transactions or generating summary reports.

- **Initial Result**: Employees save time on these tasks, freeing them to focus on higher-value activities.
- **Secondary Benefits**: The team sees immediate results, boosting morale and confidence in the change process.
- **Next Steps**: Based on the success of these initial automations, additional tasks can be automated over time, leading to a fully optimized reconciliation process.

Leadership's Role in Incremental Change

Leaders play a critical role in fostering the success of incremental change:

- **Celebrate Small Wins**: Acknowledge and celebrate each achievement, no matter how small. Recognition reinforces the value of incremental progress and motivates teams to continue.
- **Communicate the Bigger Picture**: While communicating individual success, also remind teams of how these improvements contribute to the broader goals. This helps maintain alignment and purpose.
- **Provide Support and Resources**: Ensure teams have the tools and support they need to implement and sustain incremental changes.

The Long-Term Impact of Incremental Change

While incremental changes may seem modest, their cumulative effect can be transformative. Over time, small improvements lay the foundation for a resilient, adaptive organization that thrives on continuous improvement. By focusing on what can be achieved today while keeping an eye on the future, leaders can drive sustainable success without overwhelming their teams or resources.

Celebrating Small Wins

Celebrating small wins is more than just a pat on the back; it is a powerful way to sustain momentum and build a culture of achievement. These celebrations validate the work of the team, reinforce the importance of incremental progress, and foster a sense of shared purpose. Recognition does not have to be extravagant—a thoughtful acknowledgment can leave a lasting impact on morale and motivation.

Why Celebrating Wins Matters

Celebrating wins, no matter their size, is more than just a morale booster—it is a strategic practice that fuels progress and sustains motivation. When teams feel their hard work is recognized and valued, they are more likely to stay engaged and committed to the larger objective. Taking the time to acknowledge achievements sends a powerful message: "Your efforts matter, and we're moving in the right direction."

The impact of celebration goes beyond the moment. Success has a contagious quality—it builds momentum. When teams see that their efforts are yielding results, they become energized to tackle the next challenge. It is a reminder that progress is being made and that their contributions are driving meaningful change. This forward motion is essential for maintaining focus and confidence during longer projects where results might take time to fully materialize.

Acknowledging wins also reinforces the behaviors and actions that align with organizational goals. It is a way of signaling what is valued within the organization. For example, if a team successfully adopts a new process that improves efficiency, celebrating that milestone reinforces the importance of innovation and adaptability. These moments function as guideposts, showing teams what good looks like and inspiring them to replicate those efforts.

Perhaps most importantly, celebrating successes strengthens the bonds between team members. When people come together to recognize shared achievements, it turns individual efforts into collective victories. Whether it is a simple shoutout in a meeting or a team-wide celebration, these moments of shared pride foster a sense of camaraderie. They remind everyone that they are part of something bigger—that their individual roles contribute to a greater whole.

Celebrations do not have to be extravagant to be effective. Sometimes, a heartfelt thank-you, a public acknowledgment during a team meeting, or a small gesture of appreciation like a shared lunch can make a profound difference. What matters most is the genuine recognition of effort and progress, creating a culture where accomplishments—big and small—are acknowledged and valued.

How to Celebrate Small Wins Effectively

- **Public Recognition**: Share accomplishments during team meetings, in company newsletters, or via email. Public recognition amplifies the significance of the achievement and highlights the contributions of those involved.
- **Personalized Acknowledgment**: Take the time to personally thank individuals or teams for their efforts. A simple, heartfelt "thank you" can resonate deeply and leave a lasting impression.

- **Tangible Rewards**: Small tokens of appreciation, such as gift cards, certificates, or branded items, can serve as tangible reminders of the achievement.
- **Team Gatherings**: Hosting a team lunch, coffee break, or celebratory happy hour provides an opportunity to relax and reflect on the success while strengthening team bonds.
- **Visual Reminders**: Display a progress board or achievement wall where milestones are marked. This creates a visible reminder of the team's collective progress and reinforces a sense of accomplishment.

Examples of Celebrating Small Wins

1. **Manufacturing Efficiency**: A production team implements a minor workflow adjustment that reduces downtime by 10%. The team leader arranges a quick huddle to acknowledge the effort, shares the news in the company's internal communications, and treats the team to coffee and donuts.
2. **Sales Success**: A salesperson lands a new client using a strategy developed during a recent training. Their manager sends a congratulatory email to the entire department, highlighting the achievement and sharing it as a best practice.
3. **Process Improvement**: A cross-functional team identifies and implements a small automation in the accounts payable process, saving hours of manual work each week. Leadership organizes a lunch to thank the team and invites them to share their experience during a company-wide meeting.

Incorporating Celebrations into the Culture

For celebrations to resonate beyond the moment, they must be woven into the fabric of the organization's culture. This means moving beyond ad-hoc acknowledgments and fostering an environment where recognizing success is intentional and deeply rooted in the company's values. Here's how leaders can make celebrations both meaningful and impactful.

Tie Celebrations to Core Values

Celebrations should do more than mark an achievement—they should reinforce the organization's mission and values. When leaders connect wins to these foundational principles, they provide greater meaning to the accomplishment. For example, if innovation is a core value, recognize how a successful project highlighted creative problem-solving. Similarly, if customer satisfaction is central to the company's mission, celebrate how the achievement improved the customer experience. By framing celebrations in this context, teams see their work as part of a bigger picture, fostering pride and alignment with organizational goals.

Make Celebrations Inclusive

Celebrations should involve everyone who contributed to the success, regardless of their role or level of involvement. Whether someone led the initiative or provided essential support behind the scenes, acknowledging every contributor creates a culture of inclusivity. Inclusive recognition not only boosts morale but also encourages broader participation in future initiatives. For example, during a team celebration, take time to highlight specific contributions, showing that no effort goes unnoticed. This approach builds trust and camaraderie while emphasizing the value of collective success.

Document and Share Wins

The impact of a celebration does not have to end when the event is over. Documenting successes in a shared repository creates an ongoing source of inspiration and learning for the entire organization. Teams can look back on these stories to find motivation, discover best practices, and replicate strategies that have proven effective. For instance, maintaining a “success wall” on the company intranet or a dedicated section in team newsletters allows these achievements to be showcased and celebrated widely. Sharing these stories not only reinforces a culture of appreciation but also reminds everyone that progress is possible, even in challenging times.

By embedding celebrations into the culture in thoughtful and intentional ways, leaders can ensure that recognizing success becomes a lasting and powerful motivator. Celebrations, when tied to values, made inclusive, and shared openly, transform from simple acknowledgments into catalysts for sustained engagement and excellence.

Conclusion

A solid action plan is the cornerstone of effective problem-solving, transforming Root Cause Analysis from a theoretical exercise into meaningful progress. Without one, even the most insightful RCA risks becoming an exercise in futility—leaving problems unresolved, opportunities wasted, and team morale diminished. If issues persist without resolution, employees may view RCA as an empty ritual, eroding engagement and discouraging future participation.

The purpose of an action plan is not just to map out tasks, but to provide clarity, accountability, and a sense of direction. It sets priorities that reflect organizational goals, assigns ownership to ensure progress, and breaks down complex challenges into manageable steps. A robust action plan is the bridge that connects analysis to action, ensuring that every effort contributes to long-term success.

This chapter has offered a practical framework for creating action plans that inspire teams, drive progress, and foster a culture of continuous improvement. By focusing on clear objectives, prioritization, and celebrating incremental wins, leaders can ensure their teams remain motivated and aligned. A well-crafted action plan does not just resolve the issue at hand—it creates momentum for sustained improvement and positions the organization for future success.

Chapter 12: Implementing Solutions with PDCA

Introduction

Implementation is where planning meets action. No matter how thorough the Root Cause Analysis (RCA) or how detailed the action plan, without proper execution, the effort remains theoretical. The challenge lies in ensuring that the solutions not only address the root cause but also deliver measurable, sustainable improvements. This requires a structured approach that allows for real-time adjustments and continuous refinement.

The Plan-Do-Check-Act (PDCA) cycle offers a proven framework for managing the complexity of implementation. Originating from the work of quality management pioneers like W. Edwards Deming, PDCA provides a systematic yet flexible process that ensures solutions are executed thoughtfully while enabling iterative improvement. By adopting PDCA, organizations can move from static action plans to dynamic problem-solving, achieving both immediate and long-term success.

This chapter explores the PDCA cycle in depth, guiding you through each phase—Plan, Do, Check, and Act—with practical strategies and examples. By the end, you will have a clear understanding of how to use PDCA to turn insights into impactful, measurable actions that drive lasting results.

Understanding the PDCA Cycle

The Plan-Do-Check-Act (PDCA) cycle, also known as the Deming Cycle or Shewhart Cycle, is a cornerstone of quality management and continuous improvement. It is a systematic, iterative process designed to test changes, evaluate their outcomes, and refine them based on results. Its strength lies in its simplicity, flexibility, and ability to foster continuous learning.

The Origins of PDCA

The PDCA cycle was popularized by Dr. W. Edwards Deming, a renowned quality management expert, though its roots can be traced back to Walter A. Shewhart, who first conceptualized the model in the 1920s. Shewhart introduced the idea of iterative learning and process control to improve industrial production quality. Deming later refined and expanded the concept, advocating its use not only in manufacturing but also in broader management practices.

Deming used the PDCA framework as a foundation for post-war industrial reconstruction in Japan, where it became integral to the success of companies like Toyota. Over time, the PDCA cycle has evolved into a universal tool for problem-solving, applicable across industries ranging from healthcare and education to government and technology.

What Is the PDCA Cycle?

At its core, PDCA is a structured methodology for continuous improvement. It provides a framework for identifying problems, implementing solutions, measuring their effectiveness, and making iterative adjustments. The cycle emphasizes learning through action and encourages organizations to test hypotheses, validate results, and refine their approaches in a controlled, methodical manner.

The beauty of PDCA lies in its cyclical nature—once you complete a cycle, you can begin again, building on previous learnings and striving for further improvement. This iterative process fosters a mindset of constant progress, ensuring that no solution is seen as final but as a stepping stone toward greater excellence.

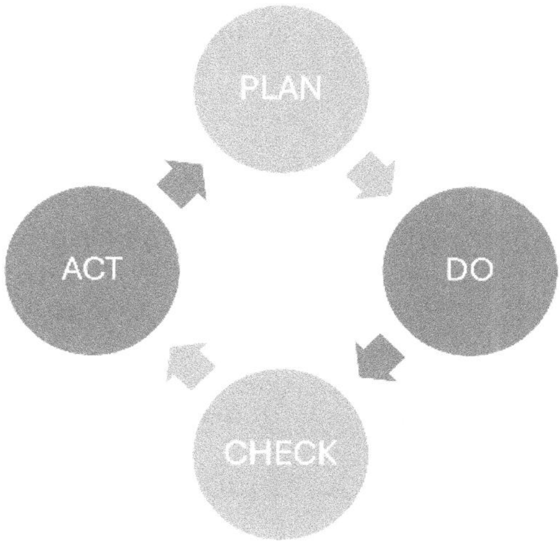

Why Use PDCA?

1. **Encourages Continuous Improvement**: Unlike one-time fixes, PDCA promotes a culture of ongoing refinement, helping organizations adapt to changes and challenges over time.
2. **Reduces Risk**: By testing changes on a small scale before full implementation, PDCA minimizes the risk of failure and allows for controlled experimentation.
3. **Builds Organizational Learning**: The cycle emphasizes documentation, analysis, and feedback, ensuring that every iteration contributes to the organization's knowledge base.

4. **Adaptable Across Industries**: Whether improving patient care in a hospital, enhancing supply chain efficiency in manufacturing, or refining service delivery in the public sector, PDCA is universally applicable.

The Structure of PDCA

PDCA consists of four phases—Plan, Do, Check, and Act—each building on the previous one. Together, they form a closed-loop system that enables teams to approach problem-solving methodically. Here is a high-level overview before we explore each phase in depth:

- **Plan**: Identify the problem, define objectives, and develop an action plan.
- **Do**: Implement the plan on a small scale, executing the tasks while observing outcomes.
- **Check**: Evaluate the results, comparing actual outcomes with expected results.
- **Act**: Refine the solution, address any shortcomings, and standardize successful practices.

Plan: Setting the Foundation

The "Plan" phase is the cornerstone of the PDCA cycle. It involves setting objectives, outlining the steps needed to achieve them, and anticipating challenges. This phase lays the groundwork for implementation by ensuring clarity, alignment, and preparedness.

Key Activities:

- **Define Objectives**: Clearly state what the implementation aims to achieve. Objectives should align with the goals set in the action plan. *Example*: If the goal is to reduce production downtime, the objective might be to implement a predictive maintenance program within three months.
- **Develop the Implementation Plan:** Break down the solution into specific tasks, timelines, and responsibilities. Use tools like Gantt charts or project management software to visualize the roadmap.

- **Anticipate Challenges**: Identify potential risks and constraints, such as resource limitations or stakeholder resistance, and develop mitigation strategies. *Example*: If training employees on new software is a potential bottleneck, plan for staggered training sessions to minimize disruption.
- **Establish Metrics**: Define how success will be measured. Metrics should be specific, measurable, and directly linked to the desired outcome. *Example*: A metric for the predictive maintenance program might be a 15% reduction in unplanned downtime within the first quarter.

Why It Matters: The "Plan" phase ensures that the implementation is intentional, well-coordinated, and aligned with organizational priorities. It reduces the likelihood of missteps and sets a clear direction for the team.

Do: Executing the Plan

The "Do" phase is where the plan comes to life. It involves executing the outlined tasks, coordinating efforts, and monitoring progress in real-time. This phase requires effective communication, adaptability, and a commitment to staying on course while addressing unexpected challenges.

Key Activities:

- **Implement the Solution**: Follow the steps outlined in the plan, ensuring each task is completed on schedule and as specified. *Example*: Begin the predictive maintenance program by training staff, installing sensors, and launching initial monitoring.
- **Communicate Effectively**: Keep all stakeholders informed of progress, challenges, and updates. Use regular check-ins and status reports to maintain alignment.
- **Monitor Progress**: Track the implementation against the timeline and metrics established in the "Plan" phase. Document any deviations and their causes. *Example*: If training sessions are delayed due to staff shortages, log the issue, and adjust the timeline accordingly.
- **Address Challenges**: Be prepared to adapt the plan as needed. Unexpected obstacles are inevitable, and flexibility is key to keeping the implementation on track.

Why It Matters: The "Do" phase is the execution engine of the PDCA cycle. It translates plans into action while providing the agility needed to address real-world complexities.

Check: Measuring and Analyzing Impact

The "Check" phase evaluates the results of the implemented solution. This phase focuses on comparing actual outcomes against the expected results, identifying gaps, and understanding what worked—and what did not.

Key Activities:

- **Collect Data**: Gather quantitative and qualitative data to assess the solution's impact. *Example*: Analyze downtime records before and after implementing the predictive maintenance program.
- **Compare Against Metrics**: Evaluate whether the predefined metrics have been met. Highlight any discrepancies between expected and actual results. *Example*: If downtime decreased by only 10% instead of the targeted 15%, investigate why.
- **Identify Root Causes of Variance**: If results fall short, determine whether the issue lies in the implementation process, the solution itself, or external factors. *Example*: A variance might stem from insufficient staff training or equipment compatibility issues.
- **Document Findings**: Record observations, insights, and lessons learned during the evaluation. This documentation will inform adjustments in the "Act" phase and guide future projects.

Why It Matters: The "Check" phase ensures that implementation efforts are purposeful and provides a basis for iterative improvement. It turns data into actionable insights.

Example: Revisiting the Quantification of Potential Solutions

In Chapter 9, we explored how to quantify potential solutions using tools like cost-benefit analyses, risk assessments, and feasibility studies. Leveraging the new performance data gathered allows you to revisit and validate those initial hypotheses. This step is crucial for understanding whether your predicted impacts align with reality.

Consider a scenario where your organization invested in a new software system to automate inventory management. In Chapter 9, the quantification process predicted that the system would reduce stock discrepancies by 30%, leading to $100,000 in annual savings. However, the data collected during the assessment phase shows a 20% reduction, resulting in $75,000 in savings. While still a success, this shortfall might suggest gaps in training or unforeseen limitations in the software's integration with other systems. These insights can inform the next cycle of improvements, ensuring that future investments yield closer-to-expected results.

Conversely, the assessment data might reveal unexpected benefits that were not initially quantified. The new software reduced employee stress by simplifying their workload, leading to improved morale and lower turnover rates. This discovery not only validates the initial decision but also enriches your understanding of the solution's broader impact, feeding back into your decision-making framework for future projects.

Act: Refining and Standardizing

The "Act" phase is where improvements are finalized, successful practices are standardized, and unresolved issues are addressed. This phase ensures that the solution is not only effective but also sustainable.

Key Activities:

- **Adjust the Plan**: Make refinements based on the findings from the "Check" phase. Address gaps, optimize processes, and resolve any lingering challenges. *Example*: If staff training was identified as a weak point, enhance the training program and extend its duration.

- **Standardize Success**: Incorporate successful practices into standard operating procedures (SOPs) and organizational workflows. *Example*: Embed the predictive maintenance program into the company's routine maintenance schedule.
- **Share Learnings**: Communicate the outcomes, insights, and adjustments to stakeholders. This fosters a culture of transparency and continuous improvement.
- **Prepare for the Next Cycle**: Recognize that PDCA is iterative. Use the insights gained to inform the next round of problem-solving and improvement efforts.

Why It Matters: The "Act" phase transforms short-term gains into long-term success. It ensures that improvements are institutionalized, and that the organization is prepared to build on them.

Example: Revisiting Solution Feasibility

A retail company introduced a new customer feedback system, expecting a 20% increase in actionable feedback submissions. However, the assessment phase revealed that submissions only increased by 5%. Data analysis showed that the submission process was too time-consuming, discouraging participation. By revisiting the feasibility assessment in Chapter 9, the company identified simpler submission tools, such as QR codes and automated surveys. After implementing these adjustments in the next PDCA cycle, actionable feedback submissions surged by 30%.

Chapter 13: Assessing Success

Introduction

A well-executed Root Cause Analysis (RCA) does not end with implementing solutions. It culminates in understanding whether those solutions deliver the results you need. This chapter is the bridge between theory and practice, tying together everything we have covered so far: defining the problem, uncovering causes, shortlisting improvement projects, creating action plans, and setting key performance indicators (KPIs).

But how do you know if you are truly moving the needle? How do you confirm that the changes made are solving the problem and driving meaningful improvement? And, just as importantly, what steps should you take if the results fall short of expectations?

In this chapter, we will explore how to assess whether your efforts are hitting the mark. By focusing on measurable outcomes, analyzing results against established KPIs, and embracing an iterative approach to continuous improvement, you will be equipped to determine success and refine your solutions. Whether it is about celebrating wins or pivoting to address gaps, this chapter will ensure you are not just solving problems—you are building a foundation for sustainable excellence.

Closing the Loop

Once your solution has been implemented, the next critical step is to evaluate its success. The effectiveness of your Root Cause Analysis (RCA) is not determined solely by whether an action plan was executed—it is about whether the changes delivered the intended outcomes. This requires systematically comparing three key components:

1. **The Initial Issue**: What was the baseline problem? This includes the symptoms, impacts, and root causes identified during the RCA process.
2. **The Projected Outcomes**: What were the specific improvements you expected to see because of the action plan? These are the Key Performance Indicators (KPIs), or goals established earlier.
3. **The New Data**: What does the current data tell you about the performance of the implemented solution?

Step 1: Establish a Baseline

Start by revisiting the original state of the problem. This includes the pre-implementation metrics you identified during the RCA process. For example:

- Defect rates
- Downtime hours
- Customer satisfaction scores
- Operational costs

Clearly documenting the initial problem and its impacts provides a benchmark for evaluating the solution's effectiveness.

Example: Before implementing a solution to reduce production downtime, you identified that the average downtime per shift was 3.5 hours, leading to significant delays and reduced output.

Step 2: Revisit Projected Outcomes

Next, align your assessment with the goals and KPIs established during the action-planning stage. What improvements did you expect, and why? These outcomes should have been clearly defined and tied to specific, measurable metrics.

Example: You projected that upgrading the maintenance scheduling system would reduce downtime to 1.5 hours per shift, translating to a 57% improvement.

Step 3: Collect and Analyze New Data

Now it is time to gather post-implementation data. Use the same metrics and methodologies as you did during the initial analysis to ensure consistency. Plot the results side by side to make comparisons easier.

Metric	Initial State	Projected Outcome	Post-Implementation Data
Average Downtime (hrs)	3.5	1.5	1.8
Defect Rate (%)	8%	4%	3.50%
Customer Complaints (per month)	30	10	12

Step 4: Assess the Gap

Compare the post-implementation results to the projected outcomes. Did the solution meet, exceed, or fall short of expectations? If there is a gap, determine whether it is significant enough to warrant further investigation.

- **If expectations were met or exceeded**: Celebrate the success, document the findings, and share lessons learned with the broader organization.

- **If expectations were not met**: Investigate why. This is a critical opportunity to apply the RCA mindset again. Did the solution address the wrong root cause? Were there unforeseen barriers to implementation? Was the solution poorly executed?

Example: If downtime only dropped to 1.8 hours instead of the projected 1.5, dig deeper to find out why. Was it due to incomplete staff training, a technical glitch in the new system, or unforeseen external factors?

Step 5: Decide What's Next

Once you have analyzed the outcomes, determine the next steps:

- If the solution was successful, consider whether additional improvements can build on this success.
- If the solution fell short, adjust your approach. This may involve refining the solution, revisiting the RCA to uncover overlooked factors, or addressing new challenges that have emerged.

Example: After identifying that inadequate staff training contributed to the gap in reduced downtime, you implement a targeted training program as the next phase of your action plan.

Why This Matters

This step is not just about evaluating results—it is about demonstrating accountability and maintaining credibility in the RCA process. If solutions do not deliver the expected results, it is crucial to show that the team is committed to continuous improvement rather than abandoning the effort. By thoroughly comparing initial issues, projected outcomes, and new data, organizations can close the loop on problem-solving and pave the way for sustained success.

Capturing Lessons for Continuous Improvement

Assessing success is not just about confirming whether a solution worked; it is an opportunity to step back, reflect, and refine the overall process. A lessons learned session serves as a structured way to capture insights—what went well, what did not, and what could be improved. This step ensures that each RCA cycle builds on the last, leading to smarter problem-solving, better collaboration, and more effective outcomes.

Why Conduct a Lessons Learned Session?

Every RCA initiative involves unique challenges, decisions, and dynamics. A lessons learned session helps:

- **Prevent Repetition of Mistakes:** By identifying pitfalls, teams can proactively avoid them in future projects.
- **Reinforce Best Practices:** Highlighting successful approaches ensures they become standard practice across the organization.
- **Foster a Culture of Continuous Improvement:** Reflecting on performance demonstrates a commitment to learning and growth, boosting team morale and engagement.

Structuring an Effective Lessons Learned Session

1. **Prepare the Groundwork** Before the session, gather relevant data and documentation from the project. This includes the initial problem statement, KPIs, action plans, and post-implementation results. These materials serve as a starting point for the discussion.

2. **Facilitate a Safe Space** To encourage open and honest feedback, ensure the session is framed as a learning exercise—not a blame game. Use neutral language to focus on processes, systems, and outcomes rather than individual performance.

For example, replace "Why did you fail to meet the deadline?" with "What obstacles prevented the deadline from being met?"

1. **Ask Key Questions** Guide the team through reflective questions that address all aspects of the project:
   - **Successes:** What worked well, and why? How can we replicate this in the future?
   - **Challenges:** What did not go as planned? Were there warning signs we missed?
   - **Outcomes:** Did the solution achieve its intended results? If not, what adjustments are needed?
   - **Collaboration:** Was the team dynamic effective? Were roles and responsibilities clear?
   - **Process:** Did the RCA methodology serve us well? How can it be improved?
2. **Encourage Participation** Ensure all voices are heard, including those from various levels of the organization. Cross-functional insights often reveal hidden perspectives that can improve outcomes.
3. **Document the Insights** Capture key takeaways from the discussion in a shared document or lessons learned repository. This ensures the insights are accessible for future teams and projects.

4. **Actionable Next Steps** Translate insights into specific actions, such as updating procedures, revising training programs, or adjusting the RCA process. Assign responsibility for each action to ensure follow-through.

Example: Lessons Learned in Action

Imagine a manufacturing team that implemented a solution to reduce equipment downtime. During the lessons learned session, the team identifies the following:

- **Success:** Automating maintenance scheduling reduced downtime by 30%, exceeding the goal.
- **Challenge:** Communication gaps during implementation caused delays in training operators on the new system.
- **Outcome:** While the downtime was reduced, customer delivery timelines were still impacted due to unanticipated bottlenecks in other areas of the workflow.

From these insights, the team decides to:

- Standardize operator training as part of future implementations.
- Introduce a cross-functional review process to identify potential bottlenecks early.

Tying Lessons Learned to Continuous Improvement

The value of lessons learned lies in action. By embedding the insights into organizational practices, teams not only improve their RCA capabilities but also foster a mindset of ongoing development. Over time, these sessions become a cornerstone of a high-performing organization, where every challenge is an opportunity to learn, adapt, and grow.

Chapter 14: Technology for Root Cause Analysis

Root Cause Analysis (RCA) has evolved significantly since its origins, with modern tools and technologies offering new ways to gather, analyze, and present data. While the foundational principles remain timeless, leveraging innovative solutions can accelerate the process, provide deeper insights, and make RCA more accessible and collaborative. For organizations operating in tech-driven environments or dealing with complex problems, integrating these tools can be a meaningful change.

A Word of Caution: Technology is a Tool, Not a Solution

While technology offers remarkable advantages, it is important to remember that tools are enablers, not substitutes for critical thinking and expertise. The judgment, creativity, and insight of a skilled facilitator are still the backbone of effective RCA. A high-tech platform can automate data collection, but it will not ask the right questions or mediate a heated debate during an RCA session.

Moreover, do not discount the "old-school" methods. A hand-drawn process map on a flipchart can sometimes spark creativity in ways a software tool cannot. Excel spreadsheets, despite their age, remain invaluable for organizing data, tracking trends, and calculating metrics. The beauty of these traditional tools lies in their simplicity and accessibility—they can adapt to any industry, budget, or team size.

The key is to find a balance. Let technology streamline repetitive tasks and provide advanced capabilities, but rely on human expertise to interpret findings, foster collaboration, and guide meaningful action.

Data Collection and Analysis Tools

Data forms the backbone of any effective RCA. Accurate data collection and analysis can pinpoint trends, anomalies, and areas of concern. Modern tools make this process faster and more precise:

- **Sensor Technology and IoT**: Internet of Things (IoT) devices and sensors can monitor equipment in real time, providing continuous data streams on performance metrics like temperature, vibration, or cycle times. For example, predictive maintenance systems use sensor data to identify potential failures before they occur.
- **Data Visualization Software**: Tools like Tableau, Power BI, or Grafana transform raw data into clear, actionable visuals. Interactive dashboards allow stakeholders to explore data trends and identify patterns quickly.
- **Big Data Analytics**: For large-scale operations, platforms like Apache Hadoop or AWS Big Data services can process massive datasets, uncovering root causes in environments with complex interdependencies.

Collaboration and Documentation Tools

Effective RCA sessions depend on seamless collaboration and clear documentation. Modern tools foster real-time input and ensure that findings are captured for ongoing reference:

- **Digital Whiteboards**: Personally, I prefer an old-fashioned white board, however there are tools like Miro or MURAL replicate the brainstorming experience of a physical whiteboard, allowing teams to collaborate on Fishbone Diagrams, 5 Whys, or process maps—even in remote settings.
- **Cloud-Based Documentation**: Platforms like Google Workspace, Microsoft Teams, or Notion enable teams to document RCA findings, share updates, and maintain a central repository accessible to all stakeholders.
- **Workflow and Task Management Software**: Tools such as Asana, Trello, or Monday.com can track action items derived from RCA sessions, assign responsibilities, and ensure deadlines are met.

Specialized RCA and Process Improvement Software

Several software solutions are purpose-built for RCA and continuous improvement initiatives. These tools streamline the RCA process, making it easier to organize findings, identify root causes, and track corrective actions:

- **Cause Mapping Tools**: Software like Sologic's Causelink or Kepner-Tregoe's Rational Process can help structure and visualize cause-and-effect relationships in a systematic way.
- **Lean and Six Sigma Platforms**: Solutions like QI Macros, Minitab or SigmaXL offer statistical tools for analyzing data within the Lean Six Sigma framework, helping quantify process variations and identify areas for improvement. I have used QI Macros for years and find it quite easy to use.
- **Incident Management Systems**: Tools like TapRooT or Enablon integrate RCA into broader incident management workflows, making them ideal for industries focused on safety or regulatory compliance.

Automation and AI in RCA

Artificial intelligence (AI) and automation are revolutionizing the way RCA is conducted by reducing human effort and increasing accuracy:

- **AI-Powered Analysis**: Platforms like IBM Watson or Salesforce Einstein use machine learning algorithms to analyze data and suggest root causes, especially in environments with complex, multivariable problems.
- **Automated Reporting**: Many tools now offer features that automate the creation of RCA reports, ensuring consistency while saving time. For example, an automated system might pull in relevant data, apply predefined templates, and generate a visual summary of findings.
- **Predictive Analytics**: Advanced AI systems can predict future failures based on historical data, allowing organizations to pre-emptively address issues before they become significant problems.

Benefits of Technology Integration in RCA

1. **Speed**: Automating data collection and analysis significantly reduces the time required to identify root causes.
2. **Accuracy**: Advanced tools minimize human error, ensuring that insights are based on reliable data.
3. **Scalability**: Technology can handle vast datasets and complex systems that would overwhelm manual approaches.
4. **Collaboration**: Cloud-based and interactive tools make RCA accessible to global teams, fostering diverse input and perspectives.

Conclusion

Incorporating modern tools and technology into RCA enhances efficiency, precision, and collaboration, enabling organizations to tackle even the most complex challenges with confidence. By staying informed about the latest innovations and thoughtfully integrating them into your RCA practices, you can unlock new levels of problem-solving potential while positioning your organization for continuous improvement in a fast-changing world.

While tools and technology can enhance the RCA process, the human element remains irreplaceable. RCA is fundamentally a collaborative effort that thrives on diverse perspectives, open communication, and shared expertise. The ability to ask the right questions, challenge assumptions, and foster a culture of trust and transparency is uniquely human. People bring creativity, intuition, and critical thinking to the table—qualities no algorithm or software can replicate. A well-facilitated RCA session empowers individuals to contribute their insights, ensuring that the root causes identified are both accurate and actionable. It is the collective effort of engaged and motivated people that turns RCA findings into meaningful, lasting change.

Chapter 15: Conclusion: Turning Knowledge into Action

Root Cause Analysis (RCA) is far more than a tool—it is a mindset, a process, and a commitment to uncovering the truth behind challenges to drive sustainable change. Over the course of this book, we have delved into every corner of RCA, from its foundational principles to the practical steps of documenting findings, implementing solutions, and assessing success. Each chapter has been crafted to provide readers with not just the knowledge, but the confidence to apply RCA in their organizations effectively.

We began by laying the groundwork, exploring the essence of RCA and the importance of a structured approach to solving problems. From there, we examined how to document the RCA process, ensuring clarity and alignment across all stakeholders. Tools such as the 5 Whys, Fishbone Diagrams, and the A3 Template offered practical frameworks for breaking down complex problems into manageable elements.

At every step, the heart of RCA lies with the people who drive it. The facilitators, who guide teams through often complex and emotionally charged discussions, play a pivotal role in ensuring neutrality, fostering trust, and maintaining focus on facts rather than opinions. Their ability to navigate group dynamics, draw out quieter voices, and manage dominant personalities is essential to the success of any RCA session.

Equally important are the participants—those who bring their unique perspectives, insights, and expertise to the table. It is their collective knowledge that uncovers hidden truths and identifies actionable opportunities. RCA thrives on collaboration, and the success of the process hinges on creating an environment where everyone feels valued, heard, and motivated to contribute.

Quantifying potential solutions and crafting action plans brought structure to the path forward, ensuring that great ideas are paired with measurable outcomes. Implementing these solutions using frameworks like PDCA and then evaluating their effectiveness tied the RCA process together, transforming insights into impactful, lasting change. But the process does not end there. It is the responsibility of the same individuals—facilitators and participants alike—to ensure that the solutions remain effective through continuous monitoring and proactive adjustments.

It is important to recognize that RCA is a skill that improves with practice. No facilitator leads a perfect session the first time, and no team executes flawlessly without experience. With every session, you will refine your approach, learn from the dynamics of your group, and gain confidence in guiding the process. Over time, what might feel awkward or challenging will become second nature, and your ability to drive impactful outcomes will grow exponentially.

RCA is an ongoing journey. It is not just about solving today's problems but creating a culture of continuous improvement. Facilitators set the tone for this culture, demonstrating the importance of transparency, collaboration, and accountability. Participants, in turn, take ownership of their roles in driving and sustaining change, ensuring that RCA becomes a natural part of how the organization operates.

As you move forward, remember that RCA is as much about the process as it is about the results. Take the time to involve the right people, embrace transparency, and remain open to learning from successes and failures alike. This is not about perfection—it is about progress.

Whether you are a seasoned leader or new to RCA, this book is a guide to turning obstacles into opportunities. With the knowledge and tools contained within, you are well-equipped to lead your teams with clarity, tackle challenges with confidence, and create a future where continuous improvement is the norm.

Your journey does not end here—it begins. Every problem solved is a step toward greater innovation, efficiency, and collaboration. RCA is your compass; use it to navigate the complexities of today's challenges and build the resilient organizations of tomorrow.

About Al Grainger

Over a distinguished two-decade career, Al Grainger has established himself as a highly sought-after expert, frequently called upon by organizations to tackle complex challenges and drive transformative improvements. With expertise as a Lean Six Sigma Black Belt, Change Management Professional, and certified Scrum Master, he is known for his ability to quickly address critical issues and lead organizations toward enhanced efficiency and effectiveness. His approach focuses on maximizing resources, enhancing productivity, and fostering rapid adaptability, positioning him as a go-to professional for resolving intricate operational problems and guiding teams through change.

Al Grainger's extensive global experience underscores why organizations trust his expertise. He has collaborated with companies of all sizes around the world, saving them countless millions by streamlining processes, implementing effective change strategies, and invigorating teams. His achievements include driving significant production improvements, leading successful process optimization initiatives, and coaching teams to strive for excellence. For instance, his efforts have resulted in substantial cost savings and enhanced operational efficiency across diverse sectors, highlighting his ability to deliver impactful results in any organizational setting.

www.ingramcontent.com/pod-product-compliance
Lightning Source LLC
LaVergne TN
LVHW022219200926
848481LV00006B/2036